AN INTRODUCTION TO
DYNAMIC METEOROLOGY

JAMES R. HOLTON
University of Washington

ACADEMIC PRESS   New York and London

A Subsidiary of Harcourt Brace Jovanovich, Publishers

ACADEMIC PRESS, INC.
111 Fifth Avenue, New York, New York 10003

United Kingdom Edition published by
ACADEMIC PRESS, INC. (LONDON) LTD.
24/28 Oval Road, London NW1

LIBRARY OF CONGRESS CATALOG CARD NUMBER: 77-187226

Second Printing, 1973

PRINTED IN THE UNITED STATES OF AMERICA

Contents

Chapter 1   Introduction

Chapter 2   The Momentum Equation

Chapter 3   Elementary Applications of the Horizontal Equations of Motion

Chapter 4   The Continuity Equation

Chapter 5   Circulation and Vorticity

Chapter 6   The Planetary Boundary Layer

Chapter 7   Diagnostic Analysis of Synoptic Scale Motions in Middle Latitudes

Chapter 8 Numerical Prediction

Chapter 9 Atmospheric Oscillations: Linear Perturbation Theory

Chapter 10 The Origin and Motion of Midlatitude Synoptic Systems

Chapter 11 The General Circulation

Chapter 12   Tropical Motion Systems

Preface

During the past decade the rapid advances in the science of dynamic meteorology made during the 1940s and 1950s have been consolidated. There now exists a reasonably coherent theory for the development of midlatitude storms, as well as for the overall general circulation of the atmosphere. Therefore, the subject can now be organized and presented in textbook form.

In this book dynamic meteorology is presented as a cohesive subject with a central unifying body of theory—namely the quasi-geostrophic system. Quasi-geostrophic theory is used to develop the principles of diagnostic analysis, numerical forecasting, baroclinic wave theory, energy transformations, and the theory of the general circulation.

Throughout the book the emphasis is on physical principles rather than mathematical elegance. It is assumed that the reader has mastered the fundamentals of classical physics, and that he has a thorough knowledge of elementary differential and integral calculus. Some use is made of vector calculus. However, in most cases the vector operations are elementary in nature so that the reader with little background in vector operations should not experience undue difficulties.

Much of the material included in this text is based on a two-quarter course sequence for seniors majoring in atmospheric sciences at the University of Washington. It would also be suitable for first-year graduate students with no previous background in meteorology.

The actual text may be divided into two main sections. The first section, consisting of Chapters 1–6, introduces those fundamentals of fluid dynamics most relevant for understanding atmospheric motions, primarily through consideration of a number of idealized types of atmospheric flow. I have found that nearly all the material in these six chapters can be presented successfully at the senior level in about 30 lectures. The second main section, consisting of Chapters 7–11, contains the core of modern dynamic meteorology with emphasis on the central unifying role of the quasi-geostrophic theory. These chapters contain more material than can be easily covered in a senior level course. However, I have attempted to arrange the material so that sections containing more advanced results can be omitted in elementary courses. On the other hand, for more intensive graduate level courses the material presented here might be supplemented by readings from original sources.

In addition to these two main sections, the book contains a concluding chapter which stands somewhat alone. In this final chapter I have attempted to review the current status of the dynamics of the tropical atmosphere. This chapter is by necessity somewhat speculative, however, the field of tropical meteorology is too important to omit entirely from a textbook on dynamic meteorology.

A short annotated list of suggested references for further reading is presented at the end of most of the chapters. I have limited the references in most cases to books and review articles which I have found to be particularly useful. No attempt has been made to provide extensive bibliographies of original sources. In the reference lists books are referred to by author and title, papers by author and date of publication. The complete references are listed in the bibliography at the end of the book.

Acknowledgments

I am indebted to Dr. R. E. Dickinson and Professor R. S. Lindzen for reading the manuscript and offering many helpful suggestions for improvements. I am also indebted to my colleague, Professor R. J. Reed, for his constant encouragement and advice.

The following publishers and organizations granted me permission to reproduce previously published figures: Academic Press, Inc., The American Meteorological Society, The American Geophysical Union, The Royal Meteorological Society, The Meteorological Society of Japan, The World Meteorological Organization, the U.S. Navy Weather Research Facility, and the Royal Society of London. Professor Dave Fultz, Dr. Jack Kornfield, and Mr. C. P. Chang have kindly provided photographs for my use.

Finally, I wish to thank Mrs. Jane Meredith for her expert typing of the manuscript.

Chapter 1 | Introduction

Atmospheric motions occur over a very broad spectrum of scales in both space and time, ranging from the random motions of individual molecules to the mean zonal circulation which involves the entire atmosphere. Dynamic meteorology is the study of those motions of the atmosphere that are either directly associated with weather phenomena or are significant elements of the general circulation. For all such motions, the atmosphere behaves as a continuous medium or fluid. The primary goal of dynamic meteorology is to apply the laws of hydrodynamics and thermodynamics to the atmosphere in order to understand and predict the weather and climate. The basic laws of fluid dynamics describe the whole spectrum of atmospheric motions, with the exception of the molecular scale. Therefore, it is not surprising that no general solutions exist for the equations embodying these laws. In order to obtain useful results the equations must be simplified and formulated in a manner which isolates the motions of interest. A systematic simplifying technique is the method called *scale analysis*.

1.1 Scale Analysis

The primary advantage of regarding the atmosphere as a continuous medium is that the physical quantities which specify the state of the atmosphere such as pressure, density, temperature, and velocity are then continuous functions of space and time. Quantities which are continuously variable in space and time are called *field variables* or simply *fields*. Pressure, density, and temperature are *scalar fields*, while velocity is a *vector field*. The field variables are continuous functions, and they are differentiable. The laws governing the motion of the atmosphere may thus be expressed as differential equations involving the field variables and their derivatives. In order to determine whether some terms in these equations are sufficiently small compared to others so that they can be neglected in approximate analyses, it is necessary to estimate not only the magnitudes of the fields themselves, but also the magnitudes of various derivatives of the field variables.

Scale analysis, or scaling, is a convenient technique for estimating the magnitudes of various terms in the governing equations for a particular type of motion. In scaling, typical expected values of the following quantities are specified: (1) the magnitudes of the field variables; (2) the amplitudes of fluctuations in the field variables; and (3) the characteristic length, depth, and time scales on which these fluctuations occur. These typical values are then used to compare the magnitudes of various terms in the governing equations. For example, a typical midlatitude synoptic cyclone might have a central surface pressure 20 mb below the mean surface pressure and a horizontal scale of 2000 km. Designating the amplitude of the horizontal pressure fluctuation by δp and the horizontal scale by L, the magnitude of the horizontal pressure gradient may be estimated as

$$\left(\frac{\partial p}{\partial x}, \frac{\partial p}{\partial y}\right) \sim \frac{\delta p}{L}$$

where $\sim$ means equality in order of magnitude. Substituting the values $\delta p = 20$ mb and $L = 2000$ km produces the scaling estimate $\delta p/L = 10^{-5}$ mb m^{-1}.

Pressure fluctuations of similar magnitudes occur in other motion systems of vastly different scale such as tornadoes, squall lines, and hurricanes. Thus, the magnitude of the horizontal pressure gradient can range over several orders of magnitude for systems of meteorological interest. Similar considerations are also valid for derivative terms involving other field variables. Therefore, the nature of the dominant terms in the governing equations is crucially dependent on the horizontal scale of the motions. In particular, for motions with horizontal scales of a few kilometers or less, terms involving the rotation of the earth are negligible, while for larger scales they become very important.

Because the character of atmospheric motions is strongly dependent on the horizontal scale, this scale provides a convenient method for classification of motion systems. In Table 1.1 examples of various types of motions are classified according to scales for the spectral region between 10^{-5} and 10^{9} cm.

In the following chapters scaling arguments will be used extensively in developing theoretical models for various types of atmospheric motions.

TABLE 1.1 *Typical Scales of Atmospheric Motion Systems*

Type of motion	Horizontal scale (cm)
Molecular mean free path	10^{-5}
Minute turbulent eddies	1–10
Small eddies	10–10^2
Dust devils	10^2–10^3
Gusts	10^3–10^4
Tornadoes	10^4
Thunderclouds	10^5
Fronts, squall lines	10^6–10^7
Hurricanes	10^7
Synoptic cyclones	10^8
Planetary waves	10^9
Atmospheric tides	10^9
Mean zonal wind	10^9

1.2 The Fundamental Forces

Newton's second law of motion states that the rate of change of momentum of an object referred to coordinates fixed in space equals the sum of all the forces acting. For atmospheric motions of meteorological interest, the forces which are of primary concern are the pressure gradient force, the gravitational force, and friction. These *fundamental* forces are the subject of the present section. If, as is the usual case, the motion is referred to a coordinate system rotating with the earth, Newton's second law may still be applied provided that certain *apparent* forces, the centrifugal force and the Coriolis force, are included among the forces acting. The nature of these apparent forces will be discussed in Section 1.3.

1.2.1 THE PRESSURE GRADIENT FORCE

We consider a differential volume element of air, $\delta V = \delta x\, \delta y\, \delta z$, centered at the point x_0, y_0, z_0 as illustrated in Fig. 1.1. Due to random molecular motions, momentum is continually imparted to the walls of the volume element by the surrounding air. This momentum transfer per unit time per unit area is just the *pressure* exerted on the walls of the volume element by the surrounding air. If the pressure at the center of the volume element is designated by p_0, then the pressure on the wall labeled A in Fig. 1.1 can be

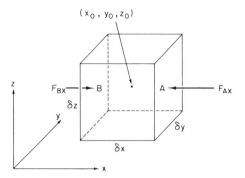

Fig. 1.1   The x component of the pressure gradient force acting on a fluid element.

expressed in a Taylor series expansion as

$$p_0 + \frac{\partial p}{\partial x} \frac{\delta x}{2} + \text{higher-order terms}$$

Neglecting the higher-order terms in this expansion, the pressure force acting on the volume element at wall A is

$$F_{Ax} = -\left(p_0 + \frac{\partial p}{\partial x} \frac{\delta x}{2}\right) \delta y\, \delta z$$

where $\delta y\, \delta z$ is the area of wall A. Similarly, the force acting on the volume element at wall B is just

$$F_{Bx} = +\left(p_0 - \frac{\partial p}{\partial x} \frac{\delta x}{2}\right) \delta y\, \delta z$$

Therefore, the net x component of the pressure force acting on the volume is

$$F_x = F_{Ax} + F_{Bx} = -\frac{\partial p}{\partial x} \delta x\, \delta y\, \delta z$$

The mass m of the differential volume element is simply the density ρ times the volume: $m = \rho\, \delta x\, \delta y\, \delta z$. Thus, the x component of the pressure gradient force per unit mass is

$$\frac{F_x}{m} = -\frac{1}{\rho} \frac{\partial p}{\partial x}$$

Similarly, it can easily be shown that the y and z components of the pressure gradient force per unit mass are

$$\frac{F_y}{m} = -\frac{1}{\rho} \frac{\partial p}{\partial y} \quad \text{and} \quad \frac{F_z}{m} = -\frac{1}{\rho} \frac{\partial p}{\partial z}$$

so that the total pressure gradient force is

$$\frac{\mathbf{F}}{m} = -\frac{1}{\rho}\,\nabla p \tag{1.1}$$

It is important to note that this force is proportional to the *gradient* of the pressure field, not to the pressure itself.

1.2.2 THE GRAVITATIONAL FORCE

Newton's law of universal gravitation states that any two elements of mass in the universe attract each other with a force proportional to their masses and inversely proportional to the square of the distance separating them. Thus, if two mass elements, M and m, are separated by a distance $r \equiv |\mathbf{r}|$ (with the vector $\mathbf{r}$ directed toward m as shown in Fig. 1.2), then the

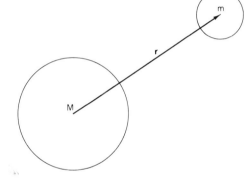

Fig. 1.2 Two spherical masses whose centers are separated by a distance **r**.

force exerted by mass M on mass m due to gravitation is

$$\mathbf{F}_g = -\frac{GMm}{r^2}\left(\frac{\mathbf{r}}{r}\right) \tag{1.2}$$

where G is a universal constant called the gravitational constant. The law of gravitation as expressed in (1.2) actually applies only to hypothetical " point " masses since for objects of finite extent $\mathbf{r}$ will vary from one part of the object to another. However, for spherically symmetric masses (1.2) may be applied if $|\mathbf{r}|$ is interpreted as the distance between the centers of the spheres. Thus, if the earth is designated as mass M and m is a mass element of the atmosphere, then the force per unit mass exerted on the atmosphere by the gravitational attraction of the earth is

$$\frac{\mathbf{F}_g}{m} \equiv \mathbf{g}^* = -\frac{GM}{r^2}\left(\frac{\mathbf{r}}{r}\right) \tag{1.3}$$

In dynamic meteorology it is customary to use as a vertical coordinate the height above mean sea level. If the mean radius of the earth is designated by a, and the distance above mean sea level is designated by z, then neglecting the small departure of the shape of the earth from sphericity, $r = a + z$. Therefore, (1.3) can be rewritten as

$$\mathbf{g}^* = \frac{\mathbf{g}_0^*}{(1 + z/a)^2} \tag{1.4}$$

where $\mathbf{g}_0^* = -(GM/a^2)(\mathbf{r}/r)$ is the value of the gravitational force at mean sea level. For meteorological applications, $z \ll a$ so that with negligible error we can let $\mathbf{g}^* = \mathbf{g}_0^*$ and simply treat the gravitational force as a constant.

1.2.3   THE FRICTION OR VISCOSITY FORCE

Although a complete discussion of the viscosity force would be rather complicated, the basic physical concept can be illustrated quite simply. We consider a layer of incompressible fluid confined between two horizontal plates separated by a distance l as shown in Fig. 1.3. The lower plate is fixed

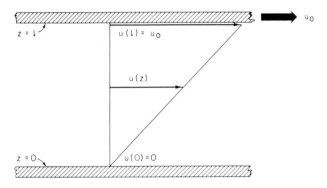

Fig. 1.3   One-dimensional steady-state viscous shear flow.

and the upper plate is moving in the x direction at a speed u_0. We find that the force tangential to the upper plate required to keep it in uniform motion is proportional to the area of the plate, the velocity, and the inverse of the distance separating the plates. Thus, we may write $F = \mu A u_0/l$, where μ is a constant of proportionality, the dynamic viscosity coefficient.

This force must just equal the force exerted by the upper plate on the fluid immediately below it. For a state of uniform motion, every horizontal layer of fluid must exert the same force on the fluid below. Therefore, taking the limit as the fluid layer depth goes to zero, we may write the viscous force per unit area, or *shearing stress*, for this special case as

$$\tau_{zx} = \mu \frac{\partial u}{\partial z}$$

where the subscripts indicate that τ_{zx} is the component of the shearing stress in the x direction due to vertical shear of the x velocity component.

From the molecular viewpoint this shearing stress results from a net downward transport of momentum by the random motion of the molecules. Because the mean x momentum increases with height, the molecules passing downward through a horizontal plane at any instant carry more momentum than those which are passing upward through the plane. Thus, there is a net downward transport of x momentum. This downward momentum transport per unit time per unit area is simply the shearing stress.

In the simple two-dimensional steady-state motion example given above there is no net viscous force acting on the elements of fluid since the shearing stress acting across the top boundary of each fluid element is just equal and opposite to that acting across the lower boundary. For the more general case of nonsteady two-dimensional shear flow in an incompressible fluid, we may calculate the net viscous force by considering again a differential volume element of fluid centered at (x_0, y_0, z_0) with sides $\delta x, \delta y, \delta z$ as shown in Fig. 1.4. If the shearing stress in the x direction acting through the center of the

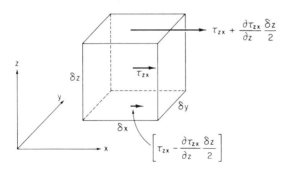

Fig. 1.4 The x component of the vertical shearing stress on a fluid element.

element is designated τ_{zx}, then the stress acting across the upper boundary may be written approximately as

$$\tau_{zx} + \frac{\partial \tau_{zx}}{\partial z} \frac{\delta z}{2}$$

while the stress acting across the lower boundary is

$$\tau_{zx} - \frac{\partial \tau_{zx}}{\partial z} \frac{\delta z}{2}$$

The net viscous force on the volume element acting in the x direction is then

$$\left(\tau_{zx} + \frac{\partial \tau_{zx}}{\partial z} \frac{\delta z}{2}\right) \delta y \, \delta x - \left(\tau_{zx} - \frac{\partial \tau_{zx}}{\partial z} \frac{\delta z}{2}\right) \delta y \, \delta x$$

so that the viscous force per unit mass due to vertical shear of the component of motion in the x direction is

$$\frac{1}{\rho} \frac{\partial \tau_{zx}}{\partial z} = \frac{1}{\rho} \frac{\partial}{\partial z} \left(\mu \frac{\partial u}{\partial z}\right) \tag{1.5}$$

For constant μ, the right-hand side of (1.5) may be simplified to $v \, \partial^2 u / \partial z^2$, where $v = \mu/\rho$ is the kinematic viscosity coefficient. For the atmosphere below 100 km, v is so small that molecular viscosity is negligible except in a thin layer within a few centimeters of the earth's surface where the vertical shear is very large. Away from this surface molecular boundary layer momentum is transferred primarily by turbulent eddy motions.

In a turbulent fluid such as the atmosphere it is often useful to picture the small-scale turbulent eddies as discrete "blobs" of fluid which move about bodily in the large-scale flow field and transfer momentum vertically in a manner analogous to the molecules in molecular viscosity. A mixing length may then be defined for the turbulent eddies, analogous to the mean free path of the molecules in molecular viscosity. By further analogy, the dissipative effects of the small-scale turbulent motions can be represented by defining an *eddy viscosity coefficient*. Thus, the simple formulation (1.5) can also be used for turbulent flow provided that the eddy viscosity coefficient is used instead of molecular viscosity. This approximation will be discussed further in Chapter 7.

1.3 Non-Newtonian Reference Frames and "Apparent" Forces

In discussing atmospheric motions it is natural to use a reference frame which is fixed with respect to the rotating earth. Newton's first law of motion states that a mass in uniform motion relative to a coordinate system fixed in space will remain in uniform motion in the absence of any forces. Such motion is referred to as *inertial motion*; and the fixed reference frame is an inertial, or newtonian, frame of reference. It is clear, however, that an object at rest with respect to the rotating earth is not at rest or in uniform motion relative to a coordinate system fixed in space. Therefore, motion which appears to be inertial motion to an observer in the rotating earth reference frame is really accelerated motion. Hence, the rotating frame is a *non-newtonian* reference frame. Newton's laws of motion can only be applied in such a frame if the acceleration of the coordinates is taken into account. The most satisfactory way of including the effects of coordinate acceleration is to introduce "apparent" forces in the statement of Newton's second law. These apparent forces are the inertial reaction terms which arise because of the coordinate acceleration. For a coordinate system in uniform rotation, two such "apparent" forces are required, the centrifugal force and the Coriolis force.

1.3.1   THE CENTRIFUGAL FORCE

We consider a ball of mass m which is attached to a string and whirled through a circle of radius r at a constant angular velocity ω. From the point of view of an observer in fixed space the speed of the ball is constant, but its direction of travel is continuously changing so that its velocity is not constant. To compute the acceleration we consider the change in velocity δV which occurs for a time increment δt during which the ball rotates through an angle

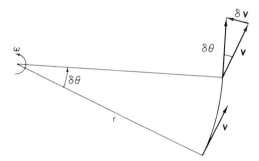

Fig. 1.5   Centripetal acceleration.

$\delta\theta$ as shown in Fig. 1.5. Since $\delta\theta$ is also the angle between the vectors $\mathbf{V}$ and $\mathbf{V} + \delta\mathbf{V}$, the magnitude of $\delta\mathbf{V}$ is just

$$|\delta V| = |V|\, \delta\theta$$

If we divide by δt and note that in the limit $\delta t \to 0$, δV is directed toward the axis of rotation, we obtain

$$\frac{d\mathbf{V}}{dt} = |\mathbf{V}|\, \frac{d\theta}{dt}\left(-\frac{\mathbf{r}}{r}\right)$$

But, $|V| = \omega r$ and $d\theta/dt = \omega$, so that

$$\frac{d\mathbf{V}}{dt} = -\omega^2 \mathbf{r} \tag{1.6}$$

Therefore, viewed from fixed coordinates the motion is one of uniform acceleration directed toward the axis of rotation, and equal to the square of

the angular velocity times the distance from the axis of rotation. This acceleration is called the *centripetal acceleration*. It is caused by the force of the string
pulling the ball.

Now suppose that we observe the motion in a coordinate system rotating
with the ball. In this rotating system the ball is *stationary*, but there is still a
force acting on the ball, namely the pull of the string. Therefore, in order to
apply Newton's second law to describe the motion relative to this rotating
coordinate system we must include an additional apparent force, the *centrifugal force*, which just balances the force of the string on the ball. Thus, the
centrifugal force is equivalent to the inertial reaction of the ball on the string,
and just equal and opposite to the centripetal acceleration.

To summarize: Observed from a fixed system the rotating ball undergoes a
constant centripetal acceleration in response to the force exerted by the string.
Observed from a system rotating along with it, the ball is stationary and the
force exerted by the string is balanced by a centrifugal force.

1.3.2 Effective Gravity

A particle of unit mass at rest on the surface of the earth, observed in a
reference frame rotating with the earth, is subject to a centrifugal force $\Omega^2\mathbf{R}$,
where Ω is the angular speed of rotation of the earth and $\mathbf{R}$ the distance from
the axis of rotation to the particle.

Thus, the weight of a particle of mass m at rest on the earth's surface,
which is just the reaction force of the earth on the particle, will generally be
less than the gravitational force $m\mathbf{g}^*$ because the centrifugal force partly
balances the gravitational force. It is, therefore, convenient to combine the
effects of the gravitational force and centrifugal force by defining an *effective
gravity* $\mathbf{g}$ such that

$$\mathbf{g} \equiv \mathbf{g}^* + \Omega^2\mathbf{R} \tag{1.7}$$

The gravitational acceleration is directed toward the center of the earth
whereas the centrifugal force is directed away from the axis of rotation.
Therefore, except at the poles and the equator, effective gravity is not directed
toward the center of the earth. As indicated by Fig. 1.6, if the earth were a
perfect sphere, the effective gravity would have an equatorward component
parallel to the surface of the earth. The earth has adjusted to compensate for
this equatorward force component by assuming the approximate shape of a
spheroid with an equatorial "bulge" so that $\mathbf{g}$ is everywhere directed normal
to the level surface. As a consequence, the equatorial radius of the earth is
about 21 km larger than the polar radius.

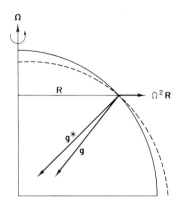

Fig. 1.6 The relationship between gravitation and effective gravity.

1.3.3 THE CORIOLIS FORCE

We have previously seen that Newton's second law may be applied in rotating coordinates to describe an object at rest with respect to the rotating system provided that an apparent force, the centrifugal force, is included among the forces acting on the object. If the object is moving with respect to the rotating system, an additional apparent force, the Coriolis force, is required in order that Newton's second law remain valid.

Suppose that an object is set in uniform motion with respect to an inertial coordinate system. If the object is observed from a rotating system with the axis of rotation perpendicular to the plane of motion, the path will appear to be curved, as indicated in Fig. 1.7. Thus, as viewed in a rotating coordinate

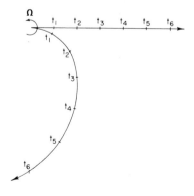

Fig. 1.7 Inertial motion as viewed from a newtonian frame (straight line) and a rotating frame (curved line).

system there is an apparent force which deflects an object in inertial motion from a straight-line path. The resulting path is curved in a direction opposite to the direction of coordinate rotation. This deflection force is the Coriolis force. Viewed from the rotating system the relative motion is an accelerated motion, with the acceleration equal to the sum of the Coriolis force and the centrifugal force. The Coriolis force, which acts perpendicular to the velocity vector, can only change the direction of travel. However, the centrifugal force, which acts radially outward, has a component along the direction of motion which increases the speed of the particle relative to the rotating coordinates as the particle spirals outward. Thus, in this example of inertial motion viewed from a rotating system the effects of both the Coriolis force and the centrifugal force are included.

Although the above example is simple enough to be readily comprehended in terms of everyday experience, it gives little insight into quantitative aspects of the Coriolis force. We now consider a particle of unit mass which is free to move on a frictionless horizontal surface on the rotating earth. If the particle is initially at rest with respect to the earth, then the only forces acting on it are the gravitational force and the apparent centrifugal force due to the rotation of the earth. As we saw in Section 1.3.2, the sum of these two forces defines the effective gravity which is directed perpendicular to the local horizontal. Suppose now that the particle is set in motion in the eastward direction by an impulsive force. Since the particle is now rotating faster than the earth, the centrifugal force on the particle will be increased. Letting Ω be the magnitude of the angular velocity of the earth, $\mathbf{R}$ the distance from the axis of rotation to the particle, and u the eastward speed of the particle relative to the ground, we may write the total centrifugal force as

$$\left(\Omega + \frac{u}{R}\right)^2 \mathbf{R} = \Omega^2 \mathbf{R} + \frac{2\Omega u \mathbf{R}}{R} + \frac{u^2 \mathbf{R}}{R^2} \tag{1.8}$$

The first term on the right is just the centrifugal force due to the rotation of the earth. This is, of course, included in the effective gravity. The other two terms represent *deflecting* forces which act outward along the vector $\mathbf{R}$ (that is, perpendicular to the axis of rotation). For synoptic scale motions $u \ll \Omega R$, and the last term may be neglected. The remaining term in (1.8), $2\Omega u(\mathbf{R}/R)$ is the *Coriolis force* due to relative motion parallel to a latitude circle. This Coriolis force can be divided into components in the vertical and meridional directions, respectively, as indicated in Fig. 1.8. Therefore, relative motion along the east–west coordinate will produce an acceleration in the north–south direction given by

$$\left(\frac{dv}{dt}\right)_{\text{Co}} = -2\Omega u \sin \phi \tag{1.9}$$

and an acceleration in the vertical given by

$$\left(\frac{dw}{dt}\right)_{\text{Co}} = 2\Omega u \cos \phi \tag{1.10}$$

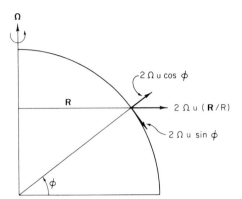

Fig. 1.8   Components of the Coriolis force due to relative motion along a latitude circle.

where ϕ is latitude and the subscript Co indicates that this is the acceleration due only to the Coriolis force. A particle moving eastward in the horizontal plane in the Northern Hemisphere will be deflected southward by the Coriolis force, whereas a westward moving particle will be deflected northward. In either case the deflection is to the right of the direction of motion. The vertical component of the Coriolis force (1.10) is ordinarily much smaller than the gravitational force so that its only effect is to cause a very minor change in the apparent *weight* of an object depending on whether the object is moving eastward or westward.

So far we have considered only the Coriolis force due to relative motion parallel to latitude circles. Suppose now that a particle initially at rest on the earth is set in motion equatorward by an impulsive force. As the particle moves equatorward it will conserve its angular momentum in the absence of torques in the east–west direction. Since the distance to the axis of rotation R increases for a particle moving equatorward, a relative westward velocity must develop if the particle is to conserve its absolute angular momentum. Thus, letting δR designate the change in the distance to the axis of rotation for a southward displacement from a latitude ϕ_0 to latitude $\phi_0 - \delta\phi$, we obtain by conservation of angular momentum

$$\Omega R^2 = \left(\Omega + \frac{\delta u}{R + \delta R}\right)(R + \delta R)^2$$

where δu is the eastward relative velocity when the particle has reached latitude $\phi_0 - \delta\phi$. Expanding the right-hand side, neglecting second-order differentials, and solving for δu, we obtain

$$\delta u = -2\Omega\delta R = +2\Omega a\delta\phi \sin \phi_0$$

where we have used $\delta R = -a\delta\phi \sin \phi_0$ (a is the radius of the earth). This

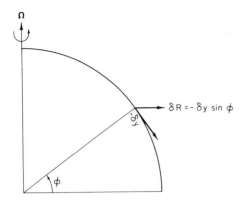

Fig. 1.9 Relationship of δR and $\delta y = a\delta\phi$ for an equatorward displacement.

relationship is illustrated in Fig. 1.9. Dividing through by the time increment δt and taking the limit as $\delta t \to 0$, we obtain from the above

$$\left(\frac{du}{dt}\right)_{Co} = 2\Omega a \frac{d\phi}{dt} \sin \phi_0 = 2\Omega v \sin \phi_0$$

where $v = a \, d\phi/dt$ is the northward velocity component.

Similarly, it is easy to show that if the particle is launched vertically at latitude ϕ_0, conservation of the absolute angular momentum will require an acceleration in the zonal direction equal to $-2\Omega w \cos \phi_0$, where w is the vertical velocity. Thus, in the general case where both horizontal and vertical relative motions are included

$$\left(\frac{du}{dt}\right)_{Co} = 2\Omega v \sin \phi - 2\Omega w \cos \phi \qquad (1.11)$$

Again, the effect of the horizontal relative velocity is to deflect the particle to the right in the Northern Hemisphere. This deflection force is negligible for motions with time scales that are very short compared to the period of the earth's rotation (a point which is illustrated by several problems at the end of the chapter). Thus, the Coriolis force is not important for the dynamics of individual cumulus clouds, but is essential to the understanding of synoptic scale systems. The Coriolis force must also be taken into account when computing long-range missile or artillery trajectories. As an example, suppose that a ballistic missile is fired due eastward at 43°N latitude. If the missile travels 1000 km at a horizontal speed $u_0 = 1000$ m sec^{-1}, by how much is the missile deflected from its eastward path by the Coriolis force? Integrating (1.9) with respect to time we find that

$$v = -2\Omega u_0 t \sin \phi \qquad (1.12)$$

where it is assumed that the deflection is sufficiently small so that we may let $u = u_0$ be constant. To find the total southward displacement we must integrate (1.12) with respect to time:

$$\int_0^{t_0} v \, dt = \int_{y_0}^{y_0 + \delta y} dy = -2\Omega u_0 \int_0^{t_0} t \, dt \sin \phi$$

Thus the total displacement is

$$\delta y = -\Omega u_0 t_0^2 \sin \phi \approx -50 \quad \text{km}$$

Therefore, the missile is deflected southward by 50 km due to the Coriolis effect. Further examples of the deflection of particles by the Coriolis force are given in some of the problems below.

1.4 Total Differentiation

Newton's second law as stated in Section 1.2 relates the acceleration experienced by any particular *parcel* of air to the sum of the forces acting on that parcel. However, for many applications in dynamic meteorology, it is necessary to know the acceleration at a fixed point in space. Similarly, the first law of thermodynamics provides an expression for the rate of change of temperature of a particular air parcel. But, what is often required is, rather, an expression for the temperature change at a fixed point. Thus, it is necessary to derive a relationship between the rate of change of a field variable following the motion and its rate of change at a fixed point. The former is called the *substantial* or *total* derivative, while the latter is called the *local* derivative (it is merely the partial derivative with respect to time).

To derive a relationship between the total derivative and the local derivative it is convenient to refer to a particular field variable, temperature, for example. Suppose that the temperature measured on a balloon which moves with the wind is T_0 at the point x_0, y_0, z_0 and time t_0. If the balloon moves to the point $x_0 + \delta x$, $y_0 + \delta y$, $z_0 + \delta z$ in a time increment δt, then the temperature change recorded on the balloon δT may be expressed in a Taylor series expansion as

$$\delta T = \left(\frac{\partial T}{\partial t}\right) \delta t + \left(\frac{\partial T}{\partial x}\right) \delta x + \left(\frac{\partial T}{\partial y}\right) \delta y + \left(\frac{\partial T}{\partial z}\right) \delta z$$

Dividing through by δt and taking the limit $\delta t \to 0$ we obtain

$$\frac{dT}{dt} = \frac{\partial T}{\partial t} + \frac{\partial T}{\partial x}\frac{dx}{dt} + \frac{\partial T}{\partial y}\frac{dy}{dt} + \frac{\partial T}{\partial z}\frac{dz}{dt} \tag{1.13}$$

where

$$\frac{dT}{dt} \equiv \lim_{\delta t \to 0} \frac{\delta T}{\delta t}$$

is the rate of change of T following the motion. If we now let

$$\frac{dx}{dt} \equiv u, \qquad \frac{dy}{dt} \equiv v, \qquad \frac{dz}{dt} \equiv w$$

then u, v, w are the velocity components in the x, y, z directions, respectively, and we may rewrite (1.13) as

$$\frac{dT}{dt} = \frac{\partial T}{\partial t} + \left(u \frac{\partial T}{\partial x} + v \frac{\partial T}{\partial y} + w \frac{\partial T}{\partial z} \right)$$

Using vector notation this expression may be rewritten as

$$\frac{\partial T}{\partial t} = \frac{dT}{dt} - \mathbf{V} \cdot \nabla T \tag{1.14}$$

where $\mathbf{V} = \mathbf{i}u + \mathbf{j}v + \mathbf{k}w$ is the velocity vector. The term $-\mathbf{V} \cdot \nabla T$ is called the temperature *advection*. It gives the contribution to the local temperature change due to the air motion. For example, if the wind is blowing from a cold region toward a warm region $-\mathbf{V} \cdot \nabla T$ will be negative (cold advection) and the advection term will contribute negatively to the local temperature change. Thus, (1.14) states that the local rate of change of temperature equals the rate of change of temperature following the motion (that is, the heating or cooling of individual air parcels) plus the advective rate of change of temperature.

The relationship given for temperature in (1.14) holds for any of the field variables. Furthermore, the total derivative can be defined following a motion field other than the actual wind field. For example, we may wish to relate the pressure change measured by a barometer on a moving ship to the local pressure change.

Example: The surface pressure decreases by 3 mb/180 km in the eastward direction. A ship steaming eastward at 10 km/hr measures a pressure fall of 1 mb/3 hr. What is the pressure change on an island which the ship is passing? If we take the x axis oriented eastward, then the local rate of change of pressure on the island is

$$\frac{\partial p}{\partial t} = \frac{dp}{dt} - u \frac{\partial p}{\partial x}$$

where dp/dt is the pressure change observed by the ship and u the velocity of the ship. Thus,

$$\frac{\partial p}{\partial t} = \frac{-1 \text{ mb}}{3 \text{ hr}} - \left(10 \frac{\text{km}}{\text{hr}}\right)\left(\frac{-3 \text{ mb}}{180 \text{ km}}\right) = -\frac{1 \text{ mb}}{6 \text{ hr}}$$

Thus, the rate of pressure fall on the island is only half the rate measured on the moving ship.

If the total derivative of a field variable is zero, then that variable is a conservative quantity following the motion. The local change is then entirely due to advection. As we shall see later, field variables that are approximately conserved following the motion play an important role in dynamic meteorology.

1.5 Atmospheric Thermodynamics

In later chapters, as we develop the fundamentals of dynamic meteorology, we will need to make frequent use of the equation of state for an ideal gas and the first law of thermodynamics.

Letting p, T, and α denote the pressure, temperature, and specific volume, respectively, we can write the equation of state for dry air as

$$p\alpha = RT \tag{1.15}$$

where R is the gas constant for dry air.

For an ideal gas, the first law of thermodynamics may be expressed in the form

$$\dot{H} \, dt = c_v \, dT + p \, d\alpha \tag{1.16}$$

where $\dot{H}$ is the rate of external heat addition per unit mass of air, and c_v is the specific heat at constant volume. This equation states that the heat added per unit mass in a time increment dt through such processes as radiation, condensation, and conduction equals the change of internal energy plus the work done by the unit mass of air on its surroundings. If there is no heat exchange between the air and its environment, $\dot{H} = 0$ and the motion is said to be *adiabatic*. For midlatitude synoptic scale systems, the external heating rate is usually small compared to the rate of change of internal energy and the rate at which work is done by the system. Thus, these motions are often considered to be approximately adiabatic.

Applying the equation of state (1.15) and noting that $c_p = c_v + R$, where c_p is the specific heat at constant pressure, we can rewrite the first law of thermodynamics as

$$\dot{H}\, dt = c_p\, dT - \alpha\, dp$$

Dividing through by T we obtain the *entropy* form of the first law

$$\frac{\dot{H}\, dt}{T} \equiv dQ = c_p \frac{dT}{T} - \frac{\alpha\, dp}{T}$$

or with the aid of the equation of state

$$\frac{\dot{H}}{T} \equiv \frac{dQ}{dt} = c_p \frac{d \ln T}{dt} - R \frac{d \ln p}{dt} \tag{1.17}$$

Equation (1.17) gives the rate of change of entropy following the motion. The entropy Q as defined by (1.17) is a field variable which depends only on the thermodynamic state of the fluid. Thus, dQ/dt is to be regarded as a total derivative. However, "heat" is not a field variable, so that the heating rate $\dot{H}$ is not to be treated as a substantial derivative. The fact that $\dot{H}/T$ yields a perfect differential is simply one statement of the second law of thermodynamics.

The entropy change dQ/dt is often expressed in terms of the change of *potential temperature* following the motion. Potential temperature is defined as the temperature which a parcel of dry air at pressure p and temperature T would have if it were expanded or compressed dry adiabatically to a pressure of 1000 mb. The relationship between pressure and temperature for an adiabatic expansion is given by setting $dQ/dt = 0$ in (1.17). The resulting expression

$$c_p\, d \ln T = R\, d \ln p$$

can be integrated to give

$$\theta = T \left(\frac{1000}{p}\right)^{R/c_p} \tag{1.18}$$

where p is the pressure in millibars and θ is just the potential temperature as defined above. Taking the logarithm of (1.18) and differentiating, we find that

$$c_p \frac{d \ln \theta}{dt} = c_p \frac{d \ln T}{dt} - R \frac{d \ln p}{dt} \tag{1.19}$$

Comparing (1.19) and (1.17) we obtain

$$c_p \frac{d \ln \theta}{dt} = \frac{dQ}{dt} \tag{1.20}$$

Thus, potential temperature is proportional to entropy. Equation (1.20) is the form of the first law of thermodynamics which we will refer to most frequently in later chapters.

Problems

1. Assuming a spherical earth, calculate the angle between the direction of the gravitational force and the effective gravity as a function of latitude at the surface of the earth. What is the maximum value of this angle?

2. Calculate the altitude above the equator at which the effective gravity becomes equal to zero. What would be the period of rotation for an earth satellite placed into orbit at that altitude?

3. If a baseball player throws a ball a horizontal distance of 100 m at $30°$ latitude in 4 sec, by how much is it deflected horizontally as a result of the rotation of the earth?

4. Two balls 4 cm in diameter are placed 100 m apart on a frictionless horizontal plane at $43°$N latitude. If the balls are impulsively propelled directly at each other with equal speeds, at what speed must they travel so that they just miss each other?

5. A locomotive of 2×10^8 gm mass travels 50 m sec^{-1} along a straight horizontal track at $43°$N. What lateral force is exerted on the rails? Compare upward reaction force exerted by the rails for cases where the locomotive is traveling eastward and westward respectively.

6. Find the horizontal displacement of a body dropped from a fixed platform at a height h at the equator neglecting the effects of air resistance. What is the numerical value of the displacement for $h = 5$ km?

7. A bullet is fired vertically upward with initial speed w_0 at latitude ϕ. Neglecting air resistance, by what distance will it be displaced horizontally when it returns to the ground? (Neglect $2\Omega u \cos \phi$ compared to g in the vertical equation.)

8. A block of mass $M = 1$ kgm is suspended from the end of a weightless string. The other end of the string is passed through a small hole in a horizontal platform and a ball of mass $m = 10$ kgm is attached. At what angular velocity must the ball rotate on the horizontal platform to balance the weight of the block if the horizontal distance of the ball from the hole is 1 m? While the ball is rotating, the block is pulled down 10 cm. What is the new angular velocity of the ball? How much work is done in pulling down the block?

9. A particle is free to slide on a horizontal frictionless plane located at a latitude ϕ on the earth. Find the equation governing the path of the particle if it is given an impulsive eastward velocity $u = u_0$ at $t = 0$. Give the solution for the position of the particle as a function of time.

Suggested References

Kittel *et al.*, *Mechanics* (Vol. 1 of the Berkeley physics course) has excellent discussions of gravitation and noninertial reference frames at the introductory level.

Feynman *et al.*, *The Feynman Lectures on Physics*, Vol. 1 also covers gravitation and non-inertial reference frames at the introductory level in an appealing informal style.

Hess, *Introduction to Theoretical Meteorology*, contains an extensive discussion of atmospheric thermodynamics for both dry and moist atmospheres.

Haltiner and Martin, *Dynamical and Physical Meteorology*, also gives a more complete account of the principles of thermodynamics.

2 | The Momentum Equation

The mathematical statement of Newton's second law of motion is called the momentum equation. Since momentum is a vector quantity, the momentum equation is a vectorial equation. In the present chapter we transform the momentum equation from an inertial reference frame to a frame rotating with the earth. We then expand the resulting equation into its components in spherical coordinates and show how scale analysis can be used to simplify the component equations for meteorological problems.

2.1 Total Differentiation of a Vector in a Rotating System

The transformation of the momentum equation to a rotating coordinate system requires a relationship between the total derivative of a vector in an inertial reference frame and the corresponding total derivative in a rotating system.

To derive this relationship we let $\mathbf{A}$ be an arbitrary vector whose cartesian components in an inertial frame are given by

$$\mathbf{A} = \mathbf{i}A_x + \mathbf{j}A_y + \mathbf{k}A_z$$

and whose components in a frame rotating with an angular velocity $\boldsymbol{\Omega}$ are

$$\mathbf{A} = \mathbf{i}'A_x' + \mathbf{j}'A_y' + \mathbf{k}'A_z'$$

Letting $d_a \mathbf{A}/dt$ be the total derivative of $\mathbf{A}$ in the inertial frame we can write

$$\frac{d_a \mathbf{A}}{dt} = \mathbf{i} \frac{dA_x}{dt} + \mathbf{j} \frac{dA_y}{dt} + \mathbf{k} \frac{dA_z}{dt}$$

$$= \mathbf{i}' \frac{dA_x'}{dt} + \mathbf{j}' \frac{dA_y'}{dt} + \mathbf{k}' \frac{dA_z'}{dt} + \frac{d\mathbf{i}'}{dt} A_x' + \frac{d\mathbf{j}'}{dt} A_y' + \frac{d\mathbf{k}'}{dt} A_z' \qquad (2.1)$$

Now,

$$\mathbf{i}' \frac{dA_x'}{dt} + \mathbf{j}' \frac{dA_y'}{dt} + \mathbf{k}' \frac{dA_z'}{dt} \equiv \frac{d_r \mathbf{A}}{dt}$$

is just the total derivative of $\mathbf{A}$ as viewed in the rotating coordinates (that is, the rate of change of $\mathbf{A}$ following the relative motion). Furthermore, since $\mathbf{i}'$ may be regarded as a position vector of unit length, $d\mathbf{i}'/dt$ is the velocity of $\mathbf{i}'$ due to its rotation. Thus $d\mathbf{i}'/dt = \boldsymbol{\Omega} \times \mathbf{i}'$ and similarly $d\mathbf{j}'/dt = \boldsymbol{\Omega} \times \mathbf{j}'$ and $d\mathbf{k}'/dt = \boldsymbol{\Omega} \times \mathbf{k}'$. Therefore (2.1) can be rewritten as

$$\frac{d_a \mathbf{A}}{dt} = \frac{d_r \mathbf{A}}{dt} + \boldsymbol{\Omega} \times \mathbf{A} \qquad (2.2)$$

which is the desired relationship.

2.2   The Vectorial Form of the Momentum Equation in Rotating Coordinates

In an inertial reference frame Newton's second law of motion may be written symbolically as

$$\frac{d_a \mathbf{V}_a}{dt} = \sum \mathbf{F} \qquad (2.3)$$

The left-hand side represents the rate of change of the absolute velocity $\mathbf{V}_a$, following the motion as viewed in an inertial system. The right-hand side represents the sum of the real forces acting *per unit mass*. In Section 1.3 we found through simple physical reasoning that when the motion is viewed in a rotating coordinate system certain additional apparent forces must be included if Newton's second law is to be valid. The same result may be obtained by a formal transformation of coordinates in (2.3).

In order to transform this expression to rotating coordinates we must first find a relationship between $\mathbf{V}_a$ and the velocity relative to the rotating system, which we will designate by $\mathbf{V}$. This relationship is obtained by applying (2.2) to the position vector $\mathbf{r}$ for an air parcel on the rotating earth:

$$\frac{d_a \mathbf{r}}{dt} = \frac{d\mathbf{r}}{dt} + \boldsymbol{\Omega} \times \mathbf{r} \qquad (2.4)$$

But $d_a\mathbf{r}/dt \equiv \mathbf{V}_a$ and $d\mathbf{r}/dt \equiv \mathbf{V}$, therefore (2.4) may be written as

$$\mathbf{V}_a = \mathbf{V} + \mathbf{\Omega} \times \mathbf{r} \tag{2.5}$$

which states simply that the absolute velocity of an object on the rotating earth is equal to its velocity relative to the earth plus the velocity due to the rotation of the earth.

Next we apply (2.2) to the velocity vector $\mathbf{V}_a$ and obtain

$$\frac{d_a\mathbf{V}_a}{dt} = \frac{d_r\mathbf{V}_a}{dt} + \mathbf{\Omega} \times \mathbf{V}_a \tag{2.6}$$

Substituting from (2.5) into the right-hand side of (2.6) gives

$$\frac{d_a\mathbf{V}_a}{dt} = \frac{d_r}{dt}(\mathbf{V} + \mathbf{\Omega} \times \mathbf{r}) + \mathbf{\Omega} \times (\mathbf{V} + \mathbf{\Omega} \times \mathbf{r})$$

$$= \frac{d_r\mathbf{V}}{dt} + 2\mathbf{\Omega} \times \mathbf{V} - \Omega^2\mathbf{R} \tag{2.7}$$

Here $\mathbf{R}$ is a vector perpendicular to the axis of rotation, with magnitude equal to the distance to the axis of rotation, so that with the aid of a vector identity,

$$\mathbf{\Omega} \times (\mathbf{\Omega} \times \mathbf{r}) = \mathbf{\Omega} \times (\mathbf{\Omega} \times \mathbf{R}) \equiv -\Omega^2\mathbf{R}$$

Eq. (2.7) states that the acceleration following the motion in an inertial system equals the acceleration following the relative motion in a rotating system, plus the Coriolis acceleration, plus the centripetal acceleration.

If we assume that the only real forces acting on the atmosphere are the pressure gradient force, gravitation, and friction, we can rewrite Newton's second law (2.3) with the aid of (2.7) as

$$\frac{d\mathbf{V}}{dt} = -2\mathbf{\Omega} \times \mathbf{V} - \frac{1}{\rho}\nabla p + \mathbf{g} + \mathbf{Fr} \tag{2.8}$$

where $\mathbf{Fr}$ designates the friction force, and the centripetal force has been combined with gravitation in the effective gravity term $\mathbf{g}$ (see Section 1.3.2). Equation (2.8) is the statement of Newton's second law for motion relative to a rotating coordinate frame. It states that the acceleration following the relative motion in the rotating frame equals the sum of the Coriolis force, the pressure gradient force, effective gravity, and friction. It is this form of the momentum equation which is basic to most work in dynamic meteorology.

2.3 The Component Equations in Spherical Coordinates

For purposes of theoretical analysis and numerical prediction, it is necessary to expand the vectorial momentum equation (2.8) into its scalar components. Since the departure of the shape of the earth from sphericity is entirely negligible for meteorological purposes, it is convenient to expand (2.8) in spherical coordinates so that the (level) surface of the earth corresponds to a coordinate surface. The coordinate axes are then (λ, ϕ, z), where λ is longitude, ϕ is latitude, and z is the vertical distance above the surface of the earth. If the unit vectors $\mathbf{i}$, $\mathbf{j}$, $\mathbf{k}$ are now taken to be directed eastward, northward, and upward, respectively, the relative velocity becomes

$$\mathbf{V} \equiv \mathbf{i}u + \mathbf{j}v + \mathbf{k}w$$

where the components u, v, and w are defined as follows:

$$u \equiv r \cos \phi \, \frac{d\lambda}{dt}, \qquad v \equiv r \, \frac{d\phi}{dt}, \qquad w \equiv \frac{dz}{dt} \tag{2.9}$$

Here, r is the distance to the center of the earth, which is related to z by $r = a + z$, where a is the radius of the earth. Traditionally, the variable r in (2.9) is replaced by the constant a. This is a very good approximation since $z \ll a$ for the regions of the atmosphere with which the meteorologist is concerned. For notational simplicity, it is conventional to define x and y as eastward and northward distance, such that $dx = a \cos \phi \, d\lambda$ and $dy = a \, d\phi$. Thus, the horizontal velocity components are $u \equiv dx/dt$ and $v \equiv dy/dt$ in the eastward and northward directions, respectively. The (x, y, z) coordinate system defined in this way is not, however, a cartesian coordinate system because the *directions* of the $\mathbf{i}$, $\mathbf{j}$, $\mathbf{k}$ unit vectors are not constant, but are functions of position on the spherical earth. This position dependence of the unit vectors must be taken into account when the acceleration vector is expanded into its components on the sphere. Thus, we write

$$\frac{d\mathbf{V}}{dt} = \mathbf{i} \, \frac{du}{dt} + \mathbf{j} \, \frac{dv}{dt} + \mathbf{k} \, \frac{dw}{dt} + u \, \frac{d\mathbf{i}}{dt} + v \, \frac{d\mathbf{j}}{dt} + w \, \frac{d\mathbf{k}}{dt} \tag{2.10}$$

In order to obtain the component equations, it is necessary first to evaluate the rates of change of the unit vectors following the motion.

We first consider $d\mathbf{i}/dt$. Expanding the total derivative as in (1.13) and noting that $\mathbf{i}$ is a function only of x (that is, an eastward-directed vector does not change its orientation if the motion is in the north–south or vertical directions), we get

$$\frac{d\mathbf{i}}{dt} = u \, \frac{\partial \mathbf{i}}{\partial x}$$

From Fig. 2.1 we see that

$$\lim_{\delta x \to 0} \frac{|\delta \mathbf{i}|}{\delta x} = \left| \frac{\partial \mathbf{i}}{\partial x} \right| = \frac{1}{a \cos \phi}$$

and that the vector $\partial \mathbf{i}/\partial x$ is directed toward the axis of rotation. Thus, as illustrated in Fig. 2.2

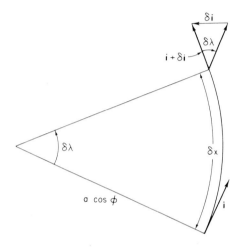

Fig. 2.1 The longitudinal dependence of the unit vector **i**.

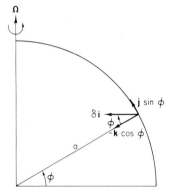

Fig. 2.2 Resolution of $\delta \mathbf{i}$ into northward and vertical components.

$$\frac{\partial \mathbf{i}}{\partial x} = \frac{1}{a \cos \phi} (\mathbf{j} \sin \phi - \mathbf{k} \cos \phi)$$

Therefore,

$$\frac{d\mathbf{i}}{dt} = \frac{u}{a \cos \phi} (\mathbf{j} \sin \phi - \mathbf{k} \cos \phi) \tag{2.11}$$

Considering now $d\mathbf{j}/dt$, we note that $\mathbf{j}$ is a function only of x and y. Thus, with the aid of Fig. 2.3 we see that $|\delta\mathbf{j}| = \delta x/(a/\tan\phi)$. Since the vector $\partial\mathbf{j}/\partial x$ is directed in the negative x direction, we have then

$$\frac{\partial\mathbf{j}}{\partial x} = -\frac{\tan\phi}{a}\mathbf{i}$$

From Fig. 2.4 it is clear that $\delta\mathbf{j} = \delta\phi/(a\,\delta\phi)$ so that

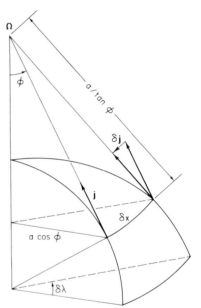

Fig. 2.3 The dependence of unit vector **j** on longitude.

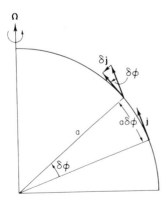

Fig. 2.4 The dependence of unit vector **j** on latitude.

$$\frac{\partial \mathbf{j}}{\partial y} = -\frac{\mathbf{k}}{a}$$

Hence,

$$\frac{d\mathbf{j}}{dt} = \frac{u \tan \phi}{a} \mathbf{i} - \frac{v}{a} \mathbf{k} \qquad (2.12)$$

Finally, by similar arguments it can be shown that

$$\frac{d\mathbf{k}}{dt} = \mathbf{i}\frac{u}{a} + \mathbf{j}\frac{v}{a} \qquad (2.13)$$

Substituting (2.11)–(2.13) into (2.10) and rearranging the terms, we obtain the spherical polar coordinate expansion of the acceleration following the relative motion

$$\frac{d\mathbf{V}}{dt} = \left(\frac{du}{dt} - \frac{uv \tan \phi}{a} + \frac{uw}{a}\right)\mathbf{i} + \left(\frac{dv}{dt} + \frac{u^2 \tan \phi}{a} + \frac{wv}{a}\right)\mathbf{j}$$
$$+ \left(\frac{dw}{dt} - \frac{u^2 + v^2}{a}\right)\mathbf{k} \qquad (2.14)$$

We next turn to the component expansion of the force terms in (2.8). The Coriolis force is expanded by noting that $\mathbf{\Omega}$ has no component parallel to $\mathbf{i}$, and that its components parallel to $\mathbf{j}$ and $\mathbf{k}$ are $2\Omega \cos \phi$ and $2\Omega \sin \phi$, respectively. Thus, using the definition of the vector cross product,

$$-2\mathbf{\Omega} \times \mathbf{V} = -2\Omega \begin{vmatrix} \mathbf{i} & \mathbf{j} & \mathbf{k} \\ 0 & \cos \phi & \sin \phi \\ u & v & w \end{vmatrix}$$

$$= -(2\Omega w \cos \phi - 2\Omega v \sin \phi)\mathbf{i} - 2\Omega u \sin \phi \mathbf{j} + 2\Omega u \cos \phi \mathbf{k} \qquad (2.15)$$

The pressure gradient force may be expressed as

$$\nabla p = \mathbf{i}\frac{\partial p}{\partial x} + \mathbf{j}\frac{\partial p}{\partial y} + \mathbf{k}\frac{\partial p}{\partial z} \qquad (2.16)$$

and effective gravity is conveniently represented as

$$\mathbf{g} = -g\mathbf{k} \qquad (2.17)$$

where g is a positive scalar ($g \simeq 980$ cm sec^{-2} at the earth's surface). Finally, friction is expanded in components as

$$\mathbf{Fr} = \mathbf{i}F_x + \mathbf{j}F_y + \mathbf{k}F_z \qquad (2.18)$$

Substituting (2.14)–(2.18) into the equation of motion (2.8) and equating all terms in the **i**, **j**, and **k** directions, respectively, we obtain

$$\frac{du}{dt} - \frac{uv \tan \phi}{a} + \frac{uw}{a} = -\frac{1}{\rho} \frac{\partial p}{\partial x} + 2\Omega\, v \sin \phi - 2\Omega\, w \cos \phi + F_x \quad (2.19)$$

$$\frac{dv}{dt} + \frac{u^2 \tan \phi}{a} + \frac{vw}{a} = -\frac{1}{\rho} \frac{\partial p}{\partial y} - 2\Omega\, u \sin \phi + F_y \quad\quad\quad (2.20)$$

$$\frac{dw}{dt} - \frac{u^2 + v^2}{a} = -\frac{1}{\rho} \frac{\partial p}{\partial z} - g + 2\Omega\, u \cos \phi + F_z \quad\quad\quad (2.21)$$

which are the eastward, northward, and vertical component momentum equations, respectively. The terms proportional to $1/a$ on the left-hand sides in (2.19)–(2.21) are called the *curvature* terms because they arise due to the curvature of the earth. Because they are nonlinear terms (that is, they are quadratic in the dependent variable), they are difficult to handle in theoretical analyses. Fortunately, as will be shown in the next section, the curvature terms are unimportant for midlatitude synoptic scale motions. However, even when the curvature terms are neglected (2.19)–(2.21) are still nonlinear partial differential equations as can be seen by expanding the total derivatives into their local and advective parts:

$$\frac{du}{dt} = \frac{\partial u}{\partial t} + u\, \frac{\partial u}{\partial x} + v\, \frac{\partial u}{\partial y} + w\, \frac{\partial u}{\partial z}$$

with similar expressions for dv/dt and dw/dt. In general the advective acceleration terms are comparable in magnitude to the local acceleration. It is primarily the presence of nonlinear advection processes that makes dynamic meteorology an interesting and challenging subject.

2.4 Scale Analysis of the Equations of Motion

In Section 1.1 we discussed the basic notion of scaling the equations of motion in order to determine whether some terms in the equations are negligible for motions of meteorological concern. Elimination of terms on scaling considerations not only has the advantage of simplifying the mathematics, but as we will show in later chapters the elimination of small terms in some cases has the very important property of completely eliminating, or *filtering*, an unwanted type of motion. The complete equations of motion (2.19)–(2.21) describe all types and scales of atmospheric motions. Sound waves, for example, are a perfectly valid solution to these equations. However, sound waves are of negligible importance in meteorological problems. Therefore, it will be a distinct advantage if, as turns out to be true, we can neglect the terms which lead to sound-wave-type solutions and filter out this unwanted class of motions.

In order to simplify (2.19)–(2.21) for synoptic scale motions we define the following characteristic scales of the field variables based on observed values for midlatitude synoptic systems:

$$U \sim 10^3 \text{ cm sec}^{-1} \qquad \text{Horizontal velocity scale}$$
$$W \sim 1 \text{ cm sec}^{-1} \qquad \text{Vertical velocity scale}$$
$$L \sim 10^8 \text{ cm} \qquad \text{Length scale}$$
$$H \sim 10^6 \text{ cm} \qquad \text{Depth scale}$$
$$\Delta P \sim 10^4 \text{ dynes cm}^{-2} \qquad \text{Horizontal pressure fluctuation scale}$$
$$L/U \sim 10^5 \text{ sec} \qquad \text{Time scale}$$

The time scale here is an advective time scale which is appropriate for pressure systems which move at approximately the speed of the horizontal wind, as is observed for synoptic scale motions. Thus L/U is the time required to travel a distance L at a speed U.

It should be pointed out here that the synoptic scale vertical velocity is not a directly measurable quantity. However, as we will show in Chapter 4, the magnitude of w can be deduced from knowledge of the horizontal velocity field.

We can now estimate the magnitude of each term in (2.19) and (2.20) for synoptic scale motions at a given latitude. It is convenient to consider disturbances centered at latitude $\phi_0 = 45°$, and introduce the notation

$$f_0 = 2\Omega \sin \phi_0 = 2\Omega \cos \phi_0 \simeq 10^{-4} \quad \text{sec}^{-1}$$

Table 2.1 shows the characteristic magnitude of each term in (2.19) and (2.20) based on the above scaling considerations. Frictional terms are not included

TABLE 2.1 *Scale Analysis of the Horizontal Momentum Equations*

	A	B	C	D	E	F
x-Component momentum equation	$\dfrac{du}{dt}$	$-2\Omega v \sin \phi$	$+2\Omega w \cos \phi$	$+\dfrac{uw}{a}$	$-\dfrac{uv \tan \phi}{a}$	$= -\dfrac{1}{\rho}\dfrac{\partial p}{\partial x}$
y-Component momentum equation	$\dfrac{dv}{dt}$	$+2\Omega u \sin \phi$		$+\dfrac{vw}{a}$	$+\dfrac{u^2 \tan \phi}{a}$	$= -\dfrac{1}{\rho}\dfrac{\partial p}{\partial y}$
Scales of individual terms	$\dfrac{U^2}{L}$	$f_0 U$	$f_0 W$	$\dfrac{UW}{a}$	$\dfrac{U^2}{a}$	$\dfrac{\Delta p}{\rho L}$
Magnitudes of the terms (cm sec^{-2})	10^{-2}	10^{-1}	10^{-4}	10^{-6}	10^{-3}	10^{-1}

because for the synoptic time scale frictional dissipation is thought to be of secondary importance above the lowest kilometer. The role of friction in the boundary layer near the ground will be discussed in Chapter 6.

2.4.1 THE GEOSTROPHIC APPROXIMATION

It is apparent from Table 2.1 that for midlatitude synoptic scale disturbances the Coriolis force (Term B) and the pressure gradient force (Term F) are in approximate balance. Therefore, retaining only these two terms in (2.19) and (2.20), we obtain as a first approximation the *geostrophic* relationship

$$-fv = -\frac{1}{\rho}\frac{\partial p}{\partial x} \tag{2.22}$$

$$+fu = -\frac{1}{\rho}\frac{\partial p}{\partial y} \tag{2.23}$$

where $f \equiv 2\Omega \sin \phi$ is called the Coriolis parameter. The geostrophic balance is a *diagnostic* expression which gives the approximate relationship between the pressure field and horizontal velocity in synoptic scale systems. The horizontal velocity field which satisfies (2.22) and (2.23) is called the *geostrophic wind*. Thus, given the pressure distribution at any time, it is possible to use (2.22) and (2.23) to derive the geostrophic wind, which for the scales used in Table 2.1 should approximate the actual horizontal wind to within about ten percent. However, the geostrophic equations contain no reference to time, and therefore cannot be used to predict the evolution of the velocity field. It is for this reason that the geostrophic relationship is called a diagnostic relationship.

2.4.2 APPROXIMATE PROGNOSTIC EQUATIONS; THE ROSSBY NUMBER

To obtain prediction equations it is necessary to retain the acceleration (Term A) in (2.19) and (2.20). The resulting approximate horizontal momentum equations are

$$\frac{du}{dt} - fv = -\frac{1}{\rho}\frac{\partial p}{\partial x} \tag{2.24}$$

$$\frac{dv}{dt} + fu = -\frac{1}{\rho}\frac{\partial p}{\partial y} \tag{2.25}$$

Our scale analysis showed that the acceleration terms in (2.24) and (2.25) are about an order of magnitude smaller than the Coriolis force and pressure gradient force. The fact that the horizontal flow is in approximate geostrophic equilibrium is helpful for diagnostic analysis. However, it makes actual applications of these equations in weather prognosis difficult because acceleration (which must be measured accurately) is given by the small difference between two large terms. Thus, a small error in measurement of either velocity or pressure will lead to very large errors in estimating the acceleration. This problem will be discussed in some detail in Chapter 8.

A convenient measure of the magnitude of the acceleration compared to the Coriolis force may be obtained by forming the ratio of the characteristic scales for the acceleration and the Coriolis force terms

$$\frac{U^2/L}{f_0 U}$$

This ratio is a nondimensional number called the *Rossby number* after the Swedish meteorologist C. G. Rossby, and is designated by

$$Ro \equiv \frac{U}{f_0 L}$$

Thus, the smallness of the Rossby number is a measure of the validity of the geostrophic approximation.

2.4.3 The Hydrostatic Approximation

A similar scale analysis can be applied to the vertical component of the momentum equation (2.21). Since pressure decreases by about an order of magnitude from the ground to the tropopause, the vertical pressure gradient may be scaled by P_0/H where P_0 is the surface pressure and H is the depth of the troposphere. The terms in (2.21) may then be estimated for synoptic scale motions and shown in Table 2.2. As with the horizontal component equations

TABLE 2.2 *Scale Analysis of the Vertical Momentum Equation*

z-Component momentum equation	$\dfrac{dw}{dt}$	$-2\Omega u \cos\phi$	$-\dfrac{u^2 + v^2}{a}$	$= -\dfrac{1}{\rho}\dfrac{\partial p}{\partial z}$	$-g$
Scales of individual terms	$\dfrac{UW}{L}$	$f_0 U$	$\dfrac{U^2}{a}$	$\dfrac{P_0}{\rho H}$	g
Magnitudes of the terms (cm sec^{-2})	10^{-5}	10^{-1}	10^{-3}	10^3	10^3

we consider motions centered at 45° latitude and neglect friction. The scaling indicates that to a high degree of accuracy the pressure field is in *hydrostatic equilibrium*, that is, the pressure at any point is simply equal to the weight of a unit cross-section column of air above that point.

The above analysis of the vertical momentum equation is, however, somewhat misleading. It is not sufficient to show merely that the vertical acceleration is small compared to g. Since only that part of the pressure field which varies horizontally is directly coupled to the velocity field, it is actually necessary to show that the horizontally varying pressure component is in itself in hydrostatic equilibrium with the horizontally varying density field. To do this it is convenient to first define a standard pressure, $p_0(z)$, which is the horizontally averaged pressure at each height, and a corresponding standard density, $\rho_0(z)$, defined so that $p_0(z)$ and $\rho_0(z)$ are in *exact* hydrostatic balance:

$$\frac{1}{\rho_0} \frac{dp_0}{dz} \equiv -g \tag{2.26}$$

We may then write the total pressure and density fields as

$$\begin{aligned} p(x, y, z, t) &= p_0(z) + p'(x, y, z, t) \\ \rho(x, y, z, t) &= \rho_0(z) + \rho'(x, y, z, t) \end{aligned} \tag{2.27}$$

where p' and ρ' are perturbations from the standard values of pressure and density. For an atmosphere at rest, p' and ρ' would thus be zero. Using the definitions (2.26) and (2.27) and assuming that ρ'/ρ_0 and p'/p_0 are much less than unity in magnitude we find that

$$-\frac{1}{\rho} \frac{\partial p}{\partial z} - g = -\frac{1}{(\rho_0 + \rho')} \frac{\partial}{\partial z} (p_0 + p') - g$$

$$\simeq \frac{1}{\rho_0} \left[\frac{\rho'}{\rho_0} \frac{dp_0}{dz} - \frac{\partial p'}{\partial z} \right] = -\frac{1}{\rho_0} \left[\rho' g + \frac{\partial p'}{\partial z} \right] \tag{2.28}$$

For synoptic scale motions, the terms in (2.28) have the magnitudes

$$\frac{1}{\rho_0} \frac{\partial p'}{\partial z} \sim \left[\frac{\Delta P}{\rho_0 H} \right] \sim 10 \quad \text{cm sec}^{-2}, \qquad \frac{\rho' g}{\rho_0} \sim 10 \quad \text{cm sec}^{-2}$$

Comparing these with the magnitudes of other terms in the vertical momentum equation (Table 2.2), we see that to a very good approximation the perturbation pressure field is in hydrostatic equilibrium with the perturbation density field so that

$$\frac{\partial p'}{\partial z} + \rho' g = 0 \tag{2.29}$$

Therefore, for synoptic scale motions, vertical accelerations are negligible and the vertical velocity cannot be determined from the vertical momentum equation. However, we will show in Chapter 7 that it is, nevertheless, possible to deduce the vertical motion field indirectly.

Problems

1. Derive the relationship

$$\mathbf{\Omega} \times (\mathbf{\Omega} \times \mathbf{r}) = -\Omega^2 \mathbf{R}$$

which was used in Equation (2.7).

2. Derive the expression given in Eq. (2.13) for the rate of change of **k** following the motion.

3. Compare the magnitudes of the curvature term $u^2 \tan \phi/a$ and the Coriolis force for a ballistic missile fired eastward with a velocity of 1000 m sec^{-1} at 45° latitude. If the missile travels 1000 km by how much is it deflected from its eastward path due to both these terms? Can the curvature term be neglected in this case?

4. Estimate the magnitudes of the terms in the equations of motion for a typical tornado. Use scales as follows: $U \sim 100$ m sec^{-1}, $W \sim 10$ m sec^{-1}, $L \sim 10^2$ m, $H \sim 10$ km, $\Delta p \sim 40$ mb. Is the hydrostatic approximation valid in this case?

5. Use scale analysis to determine what simplifications in the equations of motion are possible for hurricane scale disturbances. Let $U \sim 50$ m sec^{-1}, $W \sim 1$ m sec^{-1}, $L \sim 100$ km, $H \sim 10$ km, $\Delta p \sim 40$ mb. Is the hydrostatic approximation valid?

Suggested References

Hess, *Introduction to Theoretical Meteorology*, contains a derivation of the momentum equations for a rotating coordinate system which does not employ vector notation.

Batchelor, *An Introduction to Fluid Dynamics*, presents a more rigorous derivation of the momentum equation. This book provides an excellent introduction to the general subject of fluid dynamics at the beginning graduate level.

3 | Elementary Applications of the Horizontal Equations of Motion

3.1 The Geostrophic Wind

In Section 2.4 it was shown that for midlatitude synoptic scale systems the wind and pressure fields are in approximate geostrophic equilibrium, so that to a first approximation the horizontal momentum equation reduces to a diagnostic statement of geostrophic balance. In vector form this balance may be expressed as

$$\mathbf{V}_g = \mathbf{k} \times \frac{1}{\rho f} \mathbf{V}_h \, p \tag{3.1}$$

where $\mathbf{V}_g = \mathbf{i} u_g + \mathbf{j} v_g$ is the geostrophic velocity and $\mathbf{V}_h \, p$ is the horizontal pressure gradient. Thus, as indicated schematically in Fig. 3.1, the geostrophic wind is just that velocity field for which the Coriolis force exactly balances the horizontal pressure gradient force.

Isobaric Coordinates. Since in standard practice meteorological data is reported on constant pressure surfaces rather than constant height surfaces, it is usually advantageous to compute the geostrophic wind relative to surfaces of constant pressure rather than surfaces of constant height. Mathematically, this involves a transformation from z to p as the independent vertical coordinate and the expression of the horizontal pressure gradient in

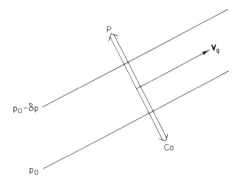

Fig. 3.1   The balance of forces for geostrophic equilibrium. The pressure gradient force is designated by P and the Coriolis force by Co.

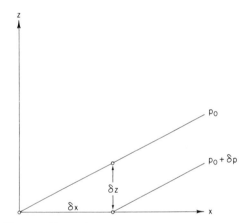

Fig. 3.2   Slope of pressure surfaces in the x, z plane.

terms of the height gradient at constant pressure. This transformation may be carried out for the x component of the pressure gradient force with the aid of Fig. 3.2. Referring to the figure we see that

$$\frac{(p_0 + \delta p) - p_0}{\delta x} = \frac{(p_0 + \delta p) - p_0}{\delta z} \frac{\delta z}{\delta x}$$

Taking the limit as $\delta x \to 0$ and $\delta z \to 0$ we obtain

$$\left(\frac{\partial p}{\partial x}\right)_z = -\left(\frac{\partial p}{\partial z}\right)_x \left(\frac{\partial z}{\partial x}\right)_p$$

which after substitution from the hydrostatic approximation may be written

$$\frac{1}{\rho}\left(\frac{\partial p}{\partial x}\right)_z = g\left(\frac{\partial z}{\partial x}\right)_p$$

where the subscripts p and z denote that pressure and height, respectively, are held constant. Substituting for the horizontal pressure gradient from (2.22), we obtain

$$fv_g = g\left(\frac{\partial z}{\partial x}\right)_p \tag{3.2}$$

as the expression for the y component of the geostrophic wind measured at constant pressure. By analogy, the x component of the geostrophic wind may be written

$$fu_g = -g\left(\frac{\partial z}{\partial y}\right)_p \tag{3.3}$$

Thus, the vectorial form of the geostrophic relationship in isobaric coordinates is

$$f\mathbf{V}_g = g\mathbf{k} \times \mathbf{V}_p z \equiv \mathbf{k} \times \mathbf{V}_p \Phi \tag{3.4}$$

where Φ is the *geopotential*, defined as the work required to raise unit mass from the surface of the earth to height z:

$$\Phi \equiv \int_0^z g \, dz \tag{3.5}$$

One advantage of isobaric coordinates is easily seen by comparing (3.1) and (3.4). In the latter equation density does not appear. Thus, a given geopotential gradient implies the same geostrophic wind at any height, whereas a given horizontal pressure gradient implies different values of the geostrophic wind depending on the density. Furthermore, if f is regarded as a constant, the horizontal divergence of the geostrophic wind at constant pressure is zero,

$$\mathbf{V}_p \cdot \mathbf{V}_g = 0$$

and the vertical component of the curl of the geostrophic wind is proportional to the horizontal Laplacian of the geopotential.

$$\mathbf{k} \cdot (\mathbf{V} \times \mathbf{V}_g) = \frac{1}{f} \mathbf{V}^2 \Phi$$

Isentropic Coordinates. It is occasionally useful to represent the geostrophic wind on surfaces of constant potential temperature. We showed in Section 1.5 that potential temperature is a measure of entropy in an ideal gas. Therefore, coordinates in which potential temperature is the independent vertical coordinate are called isentropic coordinates. Since in adiabatic frictionless flow potential temperature is conserved following the motion, isentropic coordinates are useful for tracing the actual paths of travel of individual air parcels.

Referring to Fig. 3.3 we can compare the x component of the pressure gradient computed at constant potential temperature θ to that computed at constant height

$$\frac{P_C - P_A}{\delta x} = \frac{P_C - P_B}{\delta z}\frac{\delta z}{\delta x} + \frac{P_B - P_A}{\delta x}$$

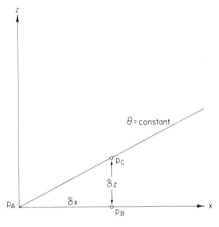

Fig. 3.3 Transformation of the geostrophic wind to isentropic coordinates.

Taking the limit as δx, $\delta z \to 0$, we obtain

$$\left(\frac{\partial p}{\partial x}\right)_\theta = \frac{\partial p}{\partial z}\left(\frac{\partial z}{\partial x}\right)_\theta + \left(\frac{\partial p}{\partial x}\right)_z$$

which after applying the hydrostatic approximation and (3.2) may be rewritten

$$fv_g = \frac{1}{\rho}\left(\frac{\partial p}{\partial x}\right)_\theta + \left(\frac{\partial \Phi}{\partial x}\right)_\theta \tag{3.6}$$

In order to eliminate density in (3.6) we recall that by definition

$$\theta \equiv T\left(\frac{1000}{p}\right)^{R/c_p}$$

where p is measured in millibars. Thus,

$$\ln\theta = \ln T - (R/c_p)\ln p + \text{constant}$$

Taking the partial derivative of this expression with respect to x on a constant θ surface yields

$$\left(\frac{\partial \ln T}{\partial x}\right)_\theta = \frac{R}{c_p}\left(\frac{\partial \ln p}{\partial x}\right)_\theta \tag{3.7}$$

Eliminating $(\partial p/\partial x)_\theta$ between (3.6) and (3.7), and applying the ideal gas law we obtain

$$fv_g = \frac{\partial}{\partial x}(c_p T + \Phi)_\theta$$

In a similar fashion it may be shown that the x component of the geostrophic wind in isentropic coordinates is

$$fu_g = -\frac{\partial}{\partial y}(c_p T + \Phi)_\theta$$

Thus, in vectorial form we have

$$\mathbf{V}_g = \frac{1}{f}\mathbf{k} \times \mathbf{V}_\theta(c_p T + \Phi) \tag{3.8}$$

Hence, the geostrophic wind in isentropic coordinates blows parallel to the isolines of the quantity $c_p T + \Phi$.

3.2 Balanced Curved Flow

The approximate horizontal momentum equations (2.24) and (2.25) may be written in vectorial form as

$$\frac{d\mathbf{V}_h}{dt} - f\mathbf{V}_h \times k = -\frac{1}{\rho}\mathbf{V}_h p \tag{3.9}$$

where $\mathbf{V}_h \equiv \mathbf{i}u + \mathbf{j}v$ is the *horizontal* velocity vector. In the following subsections we discuss the various classes of steady state (that is, time independent) flows which are governed by (3.9).

3.2.1 NATURAL COORDINATES

In order to gain an understanding of the various types of force balances possible for steady-state atmospheric flow fields, it is helpful to expand (3.9) into its components in a so-called *natural* coordinate system. The directions of the coordinates (s, n, z) in the natural coordinate system are defined by unit vectors, $\mathbf{t}$, $\mathbf{n}$, and $\mathbf{k}$ respectively; $\mathbf{t}$ is oriented parallel to the direction of flow at each point, $\mathbf{n}$ is a normal vector which is positive to the left of the flow direction, and $\mathbf{k}$ is directed vertically upward. In this system the horizontal velocity may be written $\mathbf{V_h} = V\mathbf{t}$ where $V = ds/dt$, the horizontal speed, is a nonnegative scalar. The acceleration following the motion is thus

$$\frac{d\mathbf{V_h}}{dt} = \mathbf{t}\frac{dV}{dt} + V\frac{d\mathbf{t}}{dt}$$

The rate of change of $\mathbf{t}$ following the motion may be derived from geometrical considerations with the aid of Fig. 3.4. Recalling that $|\mathbf{t}| = 1$, we see that

$$\delta\psi = \frac{\delta s}{R} = |\delta\mathbf{t}|$$

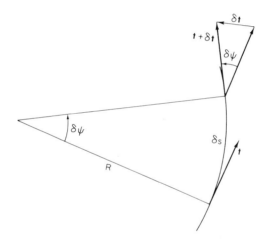

Fig. 3.4 The rate of change of the unit tangent vector $\mathbf{t}$ following the motion.

Here R is the *radius of curvature* which is taken positive when the center of curvature is in the positive $\mathbf{n}$ direction (cyclonic curvature). Noting that $\delta\mathbf{t}$ is directed parallel to $\mathbf{n}$, we find that in the limit $\delta s \to 0$

$$\frac{d\mathbf{t}}{ds} = \frac{\mathbf{n}}{R}$$

Thus,

$$\frac{d\mathbf{t}}{dt} = \frac{d\mathbf{t}}{ds}\frac{ds}{dt} = \frac{\mathbf{n}}{R}V$$

And,

$$\frac{d\mathbf{V_h}}{dt} = \mathbf{t}\frac{dV}{dt} + \mathbf{n}\frac{V^2}{R}$$

Therefore, the acceleration following the motion is the sum of the rate of change of speed of the air parcel and its centripetal acceleration due to the curvature of the trajectory. Since the Coriolis force always acts normal to the direction of motion, we can write

$$f\mathbf{V_h} \times \mathbf{k} = -fV\mathbf{n}$$

The horizontal momentum equation may thus be expanded into the following component equations in the natural coordinate system:

$$\frac{dV}{dt} = -\frac{1}{\rho}\frac{\partial p}{\partial s} \tag{3.10}$$

$$\frac{V^2}{R} + fV = -\frac{1}{\rho}\frac{\partial p}{\partial n} \tag{3.11}$$

Equations (3.10) and (3.11) express the force balances parallel to and normal to the direction of flow, respectively. For motion parallel to the isobars, $\partial p/\partial s = 0$ and the speed is constant following the motion.

3.2.2 INERTIAL FLOW

If the pressure field is horizontally uniform so that the horizontal pressure gradient vanishes, (3.11) reduces to a balance between the Coriolis force and centrifugal force:

$$\frac{V^2}{R} + fV = 0 \tag{3.12}$$

Equation (3.12) may be solved for the radius of curvature

$$R = -\frac{V}{f}$$

Since from (3.10), the speed must be constant in this case, the radius of curvature is also constant (neglecting the latitudinal dependence of f). Thus the air parcels follow circular paths in an anticyclonic (clockwise) sense. The period of this oscillation is

$$P = -\frac{2\pi R}{V} = \frac{\pi}{\Omega \sin \phi} \tag{3.13}$$

P is equivalent to the time which is required for a Foucault pendulum to turn through an angle of $180°$. Hence, P is often referred to as one-half *pendulum day*. Since both the Coriolis force and the centrifugal force due to the relative motion are caused by the inertia of the fluid, this type of motion is referred to as an inertial oscillation, and the circle of radius R is called the *inertia circle*. Pure inertial oscillations are apparently not of importance in the earth's atmosphere. However, in the oceans where velocities (and hence the radius of inertia circles) are much smaller than in the atmosphere, significant amounts of energy have been detected in currents which oscillate with the inertial period. An example of inertial oscillations in the oceans recorded by a current meter near the island of Barbados is shown in Fig. 3.5.

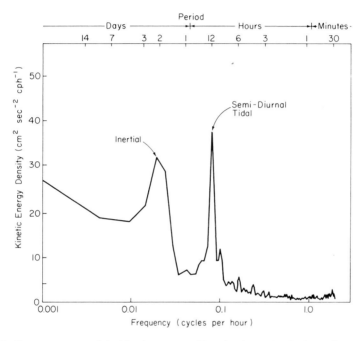

Fig. 3.5 Power spectrum of the kinetic energy at 30-m depth near Barbados (13°N latitude). This type of plot indicates the manner in which the total kinetic energy is partitioned among oscillations of different periods. Note the strong peak in the kinetic energy at 53 hr which is the period of an inertial oscillation at 13° latitude. (After Warsh, *et al.* 1971.)

3.2.3 CYCLOSTROPHIC FLOW

If the horizontal scale of a disturbance is small enough, the Coriolis force may be neglected in (3.10) compared to the pressure gradient force and the centrifugal force. The force balance normal to the direction of flow is then

$$\frac{V^2}{R} = -\frac{1}{\rho}\frac{\partial p}{\partial n}$$

If this equation is solved for V, we obtain the speed of the *cyclostrophic wind*

$$V = \left(-\frac{R}{\rho} \frac{\partial p}{\partial n} \right)^{1/2} \tag{3.14}$$

As indicated in Fig. 3.6, cyclostrophic flow may be either cyclonic or anti-

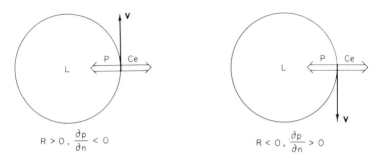

Fig. 3.6   The force balance in cyclostrophic flow, P designates the pressure gradient, Ce designates the centrifugal force.

cyclonic. In both cases the pressure gradient force is directed toward the center of curvature, and the centrifugal force away from the center of curvature.

The cyclostrophic balance approximation is valid provided that the ratio of the centrifugal force to the Coriolis force is large. This ratio V/fR is equivalent to the Rossby number discussed in Section 2.4.2. As an example of cyclostrophic scale motion we consider a typical tornado. Suppose that the tangential velocity is 30 m sec^{-1} at a distance of 300 m from the center of the vortex. Assuming that $f = 10^{-4}$ sec^{-1}, we obtain a Rossby number of

$$Ro = \frac{V}{fR} \simeq 10^3$$

Therefore, the Coriolis force can be neglected in computing the balance of forces for a mature tornado. However, the majority of tornadoes in the Northern Hemisphere are observed to rotate in a cyclonic (counterclockwise) sense. Thus, in the initial stages of development of a tornado the Coriolis force serves to deflect air parcels accelerating toward the center of low pressure to the right so that a cyclonic circulation is established about the pressure minimum. Smaller scale vortices, on the other hand, such as dust devils and water spouts, do not have a preferred direction of rotation. According to data collected by Sinclair (1965), they are observed to be anti-cyclonic as often as cyclonic.

3.2.4 GRADIENT FLOW

The steady frictionless flow which results when all terms of (3.11) are retained is called the gradient wind. Gradient flow is a three-way balance between the Coriolis force, the centrifugal force, and the pressure gradient force normal to the flow. Solution of (3.11) for V gives

$$V = -\frac{fR}{2} \pm \left(\frac{f^2R^2}{4} - \frac{R}{\rho}\frac{\partial p}{\partial n}\right)^{1/2} \tag{3.15}$$

Not all the mathematically possible roots of (3.15) correspond to physically possible solutions since it is required that V be real and nonnegative. In Table 3.1 the various roots of (3.15) are classified according to the signs of R

TABLE 3.1 *Classification of Roots of the Gradient Wind Equation*

	$R > 0$	$R < 0$
$\dfrac{\partial p}{\partial n} > 0$	Not permitted	Positive root permitted Anomalous low
$\dfrac{\partial p}{\partial n} < 0$	Positive root permitted Regular low	Positive root: anomalous high Negative root: regular high

and $\partial p/\partial n$ in order to isolate the physically meaningful solutions. The force balances for the four permitted solutions are illustrated in Fig. 3.7. Returning to Eq. (3.15), we note that in the cases of both the regular and anomalous highs the pressure gradient is limited by the requirement that the quantity under the radical be nonnegative; that is,

$$\frac{\partial p}{\partial n} < \frac{\rho R f^2}{4} \tag{3.16}$$

Thus the pressure gradient in a high must decrease linearly toward the center. It is for this reason that the pressure field near the center of a high is always flat and the wind gentle compared to the region near the center of a low.

The so-called "anomalous" gradient wind balance cases illustrated in Fig. 3.7 are unlikely to occur because as the wind field accelerates in response to an imposed pressure distribution the deflection by the Coriolis force will tend to produce a normal gradient balance. Thus, as previously argued in Section 3.2.3, the Coriolis force will tend to establish a normal cyclonic circulation about a low-pressure center. Conversely, as air parcels are accelerated outward from the center of a developing high-pressure system, the Coriolis force will gradually deflect the parcels so that they will circulate

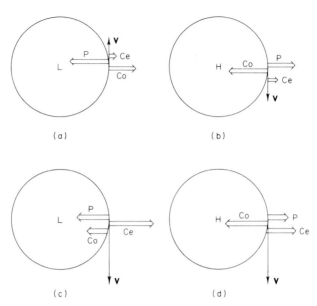

Fig. 3.7   Force balances in the four types of gradient flow: (a) regular low, (b) regular high, (c) anomalous low, (d) anomalous high.

clockwise about the high. When the motion reaches gradient equilibrium the speed will be that of a normal anticyclone.

Geostrophic versus Gradient Winds. The geostrophic wind speed in natural coordinates is given by the relationship

$$V_g \equiv -\frac{1}{\rho f}\frac{\partial p}{\partial n}$$

Using this definition, (3.11) can be rewritten as

$$\frac{V^2}{R} + fV - fV_g = 0$$

Thus the ratio of the geostrophic wind to the gradient wind is

$$\frac{V_g}{V} = 1 + \frac{V}{fR} \tag{3.17}$$

For cyclonic flow $(R > 0)$ V_g is larger than V, while for anticyclonic flow $(R < 0)$ V_g is smaller than V. Therefore, the geostrophic wind is an over-estimate of the balanced wind in a region of cyclonic curvature, and an underestimate in a region of anticyclonic curvature. For midlatitude synoptic systems, the difference between the gradient and geostrophic wind speeds rarely exceeds 10–15 percent. (Note that the magnitude of V/fR in (3.17) is just the Rossby number.) However, for the tropical cyclone scale, the Rossby number is of order unity and the gradient wind formula must be applied rather than the geostrophic wind.

3.3 Trajectories and Streamlines

In the natural coordinate system used in the previous section to discuss balanced flow the s coordinate was defined as the curve in the horizontal plane traced out by the path of an air parcel. The path followed by a particular air parcel over a finite period of time is called the *trajectory* of the parcel. Thus, the radius of curvature R of the path s referred to in the gradient wind equation is the radius of curvature for a parcel trajectory. In practice R is often estimated by using the radius of curvature of the isobars, since the latter can easily be measured at any time from a synoptic map. However, for gradient flow the isobars are actually streamlines (that is, lines which are everywhere parallel to the instantaneous wind velocity).

It is important to distinguish clearly between streamlines, which give a "snapshot" of the velocity field at any instant, and trajectories, which trace the motion of individual fluid parcels over a finite time interval. Mathematically, the horizontal trajectory is given by the integration of

$$\frac{ds}{dt} = V(x, y, t) \tag{3.18}$$

over a finite time span; whereas the streamline is given by the integration of

$$\frac{dy}{dx} = \frac{v(x, y, t_0)}{u(x, y, t_0)} \tag{3.19}$$

with respect to x for a constant time. Only for steady-state systems (in which the local rate of change of velocity is zero) will the streamlines and trajectories coincide. In order to gain an appreciation for the possible errors involved in using the curvature of the streamlines in the gradient wind equation, it is necessary to investigate the relationship between the curvature of the trajectories and the curvature of the streamlines for a moving pressure system.

If β is the angular direction of the wind and R_t and R_s designate the radii of curvature of the trajectories and streamlines, respectively, we see with the aid of Fig. 3.8 that $\delta s = R\delta\beta$ so that in the limit $\delta s \to 0$

$$\frac{d\beta}{ds} = \frac{1}{R_t} \quad \text{and} \quad \frac{\partial\beta}{\partial s} = \frac{1}{R_s} \tag{3.20}$$

where $d\beta/ds$ means the rate of change of wind direction along a trajectory

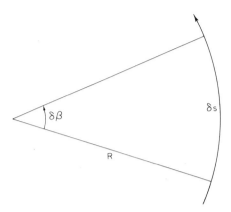

Fig. 3.8   The relationship between the change in angular direction of the wind $\delta\beta$ and the radius of curvature R.

and $\partial\beta/\partial s$ is the rate of change of wind direction along a streamline at any instant. Thus, the rate of change of wind direction following the motion is

$$\frac{d\beta}{dt} = \frac{d\beta}{ds}\frac{ds}{dt} = \frac{V}{R_t} \tag{3.21}$$

or,

$$\frac{d\beta}{dt} = \frac{\partial\beta}{\partial t} + V\frac{\partial\beta}{\partial s} = \frac{\partial\beta}{\partial t} + \frac{V}{R_s} \tag{3.22}$$

Combining (3.21) and (3.22) we obtain a formula for the local turning of the wind

$$\frac{\partial\beta}{\partial t} = V\left(\frac{1}{R_t} - \frac{1}{R_s}\right) \tag{3.23}$$

Equation (3.23) indicates that only if the trajectories and streamlines coincide, will the wind direction remain constant in time.

In general midlatitude synoptic pressure patterns move eastward as a result of advection by the upper level westerly winds. In such cases there will be local turning of the wind due to the motion of the system even if the shape of the pressure pattern does not change following its motion. For simplicity, we now consider a circular pattern of isobars moving at a constant velocity $\mathbf{C}$ without change of shape. It must be kept clearly in mind here that $\mathbf{C}$ designates the velocity at which the pressure pattern is moving and should not be confused with the wind velocity, which in this case (neglecting friction) is given by the gradient wind equation. Thus, the isobars are streamlines and the local turning of the wind is entirely due to the motion of the streamline pattern, so that

$$\frac{\partial \beta}{\partial t} = -\mathbf{C} \cdot \nabla \beta = -C \cos \gamma \frac{\partial \beta}{\partial s}$$

where γ is the angle between the streamlines (isobars) and direction of motion of the system. Substituting the above into (3.23) and solving for R_s we obtain the desired relationship between the curvature of the streamlines and the curvature of trajectories:

$$R_s = R_t \left(1 - \frac{C \cos \gamma}{V} \right) \tag{3.24}$$

Equation (3.24) can be used to compute the curvature of the trajectory anywhere on a moving pattern of streamlines. In Fig. 3.9 the curvatures of the trajectories for parcels initially located due north, east, south, and west of the

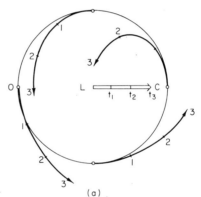

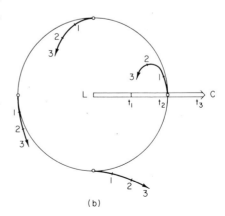

Fig. 3.9 Trajectories for moving circular circulation systems with (a) $V = 2C$ and (b) $2V = C$. Numbers indicate positions at successive times.

center of a cyclonic system are shown both for the case of a wind speed greater than the speed of movement of the isobars, and the case of a wind speed less than the speed of movement of the isobars. In these examples the plotted trajectories are based on a geostrophic balance so that the isobars are equivalent to streamlines. It is also assumed for simplicity that the wind speed does not depend on the distance from the center of the system. In the case shown in Fig. 3.9b there is actually a region where the curvature of the trajectories is opposite that of the streamlines. Since synoptic scale pressure systems usually move at a speed comparable to the wind speed, the gradient wind speed computed on the basis of the curvature of the isobars is often no better an approximation to the actual wind speed than is the geostrophic wind. In fact the actual gradient wind speed will vary along an isobar with the variation of the trajectory curvature.

3.4   Vertical Shear of the Geostrophic Wind: The Thermal Wind

The geostrophic wind must have vertical shear in the presence of a horizontal temperature gradient, as can easily be shown from simple physical considerations based on hydrostatic equilibrium. Recalling that the geostrophic wind is proportional to the horizontal pressure gradient, we see that for the case shown in Fig. 3.10 there is a geostrophic wind directed along the

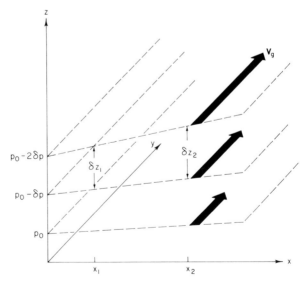

Fig. 3.10 Relationship between vertical shear of the geostrophic wind and horizontal temperature gradients.

y axis which increases in magnitude with height as the slope of the isobars increases. From the hydrostatic approximation and the ideal gas law it can be shown that the height increment δz corresponding to a given pressure increment δp is

$$\delta z \approx - \frac{\delta p}{\rho g} = - \frac{\delta p}{p} RT \qquad (3.25)$$

Therefore, referring to Fig. 3.10, since $(\delta z)_{x_1} < (\delta z)_{x_2}$ we see that $T_{x_2} > T_{x_1}$. Thus, the increase with height of the positive x-directed pressure gradient must be associated with a positive x-directed temperature gradient. The air in a vertical column at x_2 since it is warmer (less dense) must occupy a greater depth for a given pressure drop than the air at x_1.

The equations for the rate of change of the geostrophic wind components with height are most easily derived using the isobaric coordinate system. Recall that in isobaric coordinates the geostrophic wind components are

$$v_g = \frac{1}{f} \frac{\partial \Phi}{\partial x} \quad \text{and} \quad u_g = - \frac{1}{f} \frac{\partial \Phi}{\partial y} \qquad (3.26)$$

where the derivatives are evaluated with pressure held constant. Also, with the aid of the ideal gas law we can write the hydrostatic equation as

$$\frac{\partial \Phi}{\partial p} = - \alpha = - \frac{RT}{p} \qquad (3.27)$$

Differentiating (3.26) with respect to pressure and applying (3.27) we obtain

$$p \frac{\partial v_g}{\partial p} \equiv \frac{\partial v_g}{\partial \ln p} = - \frac{R}{f} \left(\frac{\partial T}{\partial x} \right)_p \qquad (3.28)$$

$$p \frac{\partial u_g}{\partial p} \equiv \frac{\partial u_g}{\partial \ln p} = \frac{R}{f} \left(\frac{\partial T}{\partial y} \right)_p \qquad (3.29)$$

or in vectorial form

$$\frac{\partial \mathbf{V}_g}{\partial \ln p} = - \frac{R}{f} \mathbf{k} \times (\nabla T)_p \qquad (3.30)$$

Equation (3.30) is often referred to as the *thermal wind* equation. However, it is actually a relationship for the vertical wind *shear* (that is, the rate of change of the geostrophic wind with respect to ln p). Strictly speaking the term thermal wind refers to the vector difference between the geostrophic winds at two levels. Designating the thermal wind vector by $\mathbf{V}_T$ we may integrate (3.30) from pressure level p_0 to level p_1 ($p_1 < p_0$) to get

$$\mathbf{V}_T = \mathbf{V}_g(p_1) - \mathbf{V}_g(p_0) = - \frac{R}{f} \int_{p_0}^{p_1} (\mathbf{k} \times \nabla T) \, d \ln p \qquad (3.31)$$

Letting $\bar{T}$ denote the mean temperature in the layer between pressure p_0 and p_1, the x and y components of the thermal wind are thus given by

$$u_T = -\frac{R}{f}\left(\frac{\partial \bar{T}}{\partial y}\right)_p \ln\left(\frac{p_0}{p_1}\right)$$

$$v_T = \frac{R}{f}\left(\frac{\partial \bar{T}}{\partial x}\right)_p \ln\left(\frac{p_0}{p_1}\right)$$

(3.32)

Alternatively, we may express the thermal wind for a given layer in terms of the horizontal gradient of the geopotential difference between the top and bottom of the layer:

$$u_T = u_g(p_1) - u_g(p_0) = -\frac{1}{f}\frac{\partial}{\partial y}(\Phi_1 - \Phi_0)$$

$$v_T = v_g(p_1) - v_g(p_0) = \frac{1}{f}\frac{\partial}{\partial x}(\Phi_1 - \Phi_0)$$

(3.33)

The equivalence of (3.32) and (3.33) can be readily verified by integrating the hydrostatic equation (3.27) vertically from p_0 to p_1 after replacing T by the mean $\bar{T}$. The result is

$$\delta\Phi \equiv \Phi_1 - \Phi_0 = R\ln\left(\frac{p_0}{p_1}\right)\bar{T}$$

(3.34)

The quantity $\delta\Phi$ is the *thickness* of the layer between p_0 and p_1 measured in units of geopotential. From (3.34) we see that the thickness is proportional to the mean temperature in the layer. Hence, lines of equal $\delta\Phi$ (isolines of thickness) are equivalent to the isotherms of mean temperature in the layer. Note also that $\delta\Phi/g$ gives the approximate thickness in units of length.

The thermal wind equation is an extremely useful diagnostic tool which is often used to check analyses of the observed wind and temperature fields for consistency. It can also be used to estimate the mean horizontal temperature advection in a layer as shown in Fig. 3.11. It is clear from the vector form of (3.33)

$$\mathbf{V}_T = \frac{1}{f}\mathbf{k} \times \nabla(\Phi_1 - \Phi_0)$$

that the thermal wind blows parallel to the isotherms (lines of constant thickness) with the warm air to the right facing downstream in the Northern hemisphere. Thus, as is illustrated in Fig. 3.11a, a geostrophic wind which turns counterclockwise with height (backs) is associated with cold-air advection. Conversely, as shown in Fig. 3.11b, clockwise turning (veering) of the geostrophic wind with height implies warm advection by the geostrophic wind

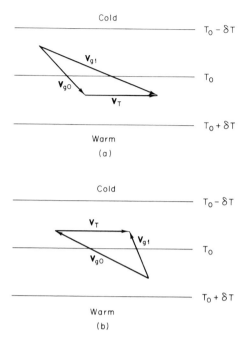

Fig. 3.11 The relationship between turning of the geostrophic wind and temperature advection: (a) backing of the wind with height, (b) veering of the wind with height.

in the layer. It is therefore possible to obtain a reasonable estimate of the horizontal temperature advection and its vertical dependence at a given location solely from data on the vertical profile of the wind given by a single sounding. Alternatively, the geostrophic wind at any level can be estimated from the mean temperature field provided that the geostrophic velocity is known at a single level. Thus, for example, if the geostrophic wind at 850 mb is known and the mean horizontal temperature gradient in the layer 850–500 mb is also known, the thermal wind equation can be applied to obtain the geostrophic wind at 500 mb.

The Barotropic Atmosphere. A barotropic atmosphere is one in which the density depends only on the pressure, $\rho \equiv \rho(p)$, so that isobaric surfaces are also surfaces of constant density. For an ideal gas, the isobaric surfaces will also be isothermal if the atmosphere is barotropic. Thus, $(\nabla T)_p = 0$ in a barotropic atmosphere, and the thermal wind equation (3.30) becomes

$$\frac{\partial \mathbf{V}_g}{\partial \ln p} = 0$$

which states that the geostrophic wind is *independent of height* in a barotropic atmosphere. Thus, barotropy provides a very strong constraint on the types of motion possible in a fluid.

The Baroclinic Atmosphere. An atmosphere in which density depends on both the temperature and the pressure, $\rho = \rho(p, T)$, is referred to as a baroclinic atmosphere. In a baroclinic atmosphere the geostrophic wind generally has vertical shear, and this shear is related to the horizontal temperature gradient by the thermal wind equation. Obviously, the baroclinic atmosphere is of primary importance in dynamic meteorology. However, as will be seen in later chapters, much can be learned by study of the simpler barotropic atmosphere.

Problems

1. An aircraft flying a course of $60°$ at air speed 200 m/sec moves relative to the ground due east ($90°$) at 225 m sec^{-1}. If the plane is flying at constant pressure, what is its rate of change in altitude assuming a steady pressure field and $f = 10^{-4}$ sec^{-1}?

2. The actual wind is directed $30°$ to the right of the geostrophic wind. If the geostrophic wind is 20 m sec^{-1}, what is the rate of change of wind speed? Let $f = 10^{-4}$ sec^{-1}.

3. A tornado rotates with constant angular velocity ω. Show that the surface pressure at the center of the tornado is given by

$$p = p_0 \exp \frac{-\omega^2 r_0{}^2}{2RT}$$

where p_0 is the surface pressure at a distance r_0 from the center and T is the temperature (assumed constant), if the temperature is $288°K$, pressure at 100 m from the center is 1000 mb, and wind speed at 100 m from center is 100 m sec^{-1}, what is the central pressure?

4. Calculate the geostrophic wind speed in meters per second for a pressure gradient of 1 mb/100 km and compare with *all* possible gradient wind speeds for the same pressure gradient and a radius of curvature of ± 500 km. Let $\rho = 10^{-3}$ gm cm^{-3} and $f = 10^{-4}$ sec^{-1}.

5. Determine the maximum possible ratio of the normal anticyclonic gradient wind speed to the geostrophic wind speed for the same pressure gradient.

6. Show that the geostrophic balance in isothermal coordinates may be written

$$f\mathbf{V}_g = \mathbf{k} \times \mathbf{V}_T(RT \ln p + \Phi)$$

7. Determine the radii of curvature for the trajectories of air parcels located 500 km to the east, north, south, and west of the center of a circular low-pressure system, respectively. The system is moving eastward at 15 m sec^{-1}. Assume geostrophic flow with a uniform tangential wind speed of 15 m sec^{-1}.

8. Determine the gradient wind speeds for the four air parcels in Problem 7 and compare these speeds with the geostrophic speed.

9. Show that as the pressure gradient approaches zero the gradient wind reduces to the geostrophic wind for a normal anticyclone and to the inertia circle for an anomalous anticyclone.

10. 1000–500 mb thickness lines are drawn at intervals of 60 m. What is the approximate corresponding mean temperature interval?

11. The mean temperature in the layer between 750 and 500 mb decreases eastward by 3°C per 100 km. If the 750 mb geostrophic wind is from the southeast at 20 m sec^{-1}, what is the geostrophic wind speed and direction at 500 mb? Let $f = 10^{-4}$ sec^{-1}.

12. What is the mean temperature advection in the 750–500 mb layer in Problem 11?

13. Suppose that a vertical column of the atmosphere is initially isothermal from 900 to 500 mb. The geostrophic wind is 10 m sec^{-1} from the south at 900 mb, 10 m sec^{-1} from the west at 700 mb, and 20 m sec^{-1} from the west at 500 mb. Calculate the mean horizontal temperature gradients in the two layers 900–700 mb and 700–500 mb. Compute the rate of advective temperature change in each layer. How long would this advection pattern have to persist in order to establish a dry adiabatic lapse rate between 600 and 800 mb? (Assume that the lapse rate is linear from 900–500 mb.)

Suggested Reference

Haltiner and Martin, *Dynamical and Physical Meteorology*, covers the material of this chapter in somewhat more detail than presented here.

Chapter

4 | The Continuity Equation

In the previous chapter we have elucidated some basic properties of horizontal flow in the atmosphere through application of a single physical principle—Newton's second law of motion. So far little has been said concerning vertical motion in the atmosphere. However, it should be clear from the scaling arguments of Chapter 2 that the vertical momentum equation cannot be used directly to deduce the vertical motion in synoptic scale systems since to a high degree of accuracy such systems are in hydrostatic balance. In the present chapter we consider an additional fundamental physical principle—the conservation of mass, which relates the horizontal and vertical motion fields. Thus, in theory the principle of mass conservation (or continuity of mass) can be used to deduce the vertical motion field from knowledge of the horizontal velocity field. However, as we shall see later, there are practical difficulties involved in applying this method.

4.1 Derivation in Cartesian Coordinates

We consider a volume element $\delta x \, \delta y \, \delta z$ which is fixed in space as shown in Fig. 4.1. For such a *fixed* volume the net rate of mass inflow through the sides must equal the rate of accumulation of mass within the volume. The rate of inflow of mass through the left-hand face per unit area is

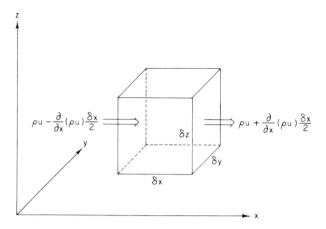

Fig. 4.1 Mass inflow into a fixed volume element due to motion along the x axis.

$$\left[\rho u - \frac{\partial}{\partial x}(\rho u)\frac{\delta x}{2}\right]$$

whereas the rate of outflow per unit area through the right-hand face is

$$\left[\rho u + \frac{\partial}{\partial x}(\rho u)\frac{\delta x}{2}\right]$$

Since the area of each of these faces is $\delta y\,\delta z$, the net rate of flow into the volume due to the x velocity component is

$$\left[\rho u - \frac{\partial}{\partial x}(\rho u)\frac{\delta x}{2}\right]\delta y\,\delta z - \left[\rho u + \frac{\partial}{\partial x}(\rho u)\frac{\delta x}{2}\right]\delta y\,\delta z = -\frac{\partial}{\partial x}(\rho u)\,\delta y\,\delta z$$

Similar expressions obviously hold for the y and z directions. Thus, the net rate of mass inflow is

$$-\left[\frac{\partial}{\partial x}(\rho u) + \frac{\partial}{\partial y}(\rho v) + \frac{\partial}{\partial z}(\rho w)\right]\delta x\,\delta y\,\delta z$$

And the mass inflow per unit volume is just $-\mathbf{V}\cdot(\rho\mathbf{V})$, which must equal the rate of mass increase per unit volume. Now the increase of mass per unit volume is just the local density change $\partial\rho/\partial t$. Therefore,

$$\frac{\partial\rho}{\partial t} + \mathbf{V}\cdot(\rho\mathbf{V}) = 0 \qquad (4.1)$$

Equation (4.1) is the mass divergence form of the continuity equation.

An alternative form of the continuity equation is obtained by applying the vector identity

$$\mathbf{V}\cdot(\rho\mathbf{V}) \equiv \rho\mathbf{V}\cdot\mathbf{V} + \mathbf{V}\cdot\mathbf{V}\rho$$

to get

$$-\frac{1}{\rho}\frac{d\rho}{dt} + \mathbf{V}\cdot\mathbf{V} = 0 \tag{4.2}$$

Equation (4.2) is the velocity divergence form of the continuity equation. It states that the fractional rate of increase of the density *following the motion* of an air parcel is equal to minus the velocity divergence. This should be clearly distinguished from (4.1) which states that the *local* rate of change of density is equal to minus the mass divergence.

Equation (4.2) may also be derived simply in a manner which illustrates the physical meaning of the velocity divergence. We consider an air parcel of mass $\delta M = \rho\,\delta A\,\delta z$ where $\delta A = \delta x\,\delta y$ is the cross-section area and δz the height of the air parcel. Now the mass of the parcel must be conserved following its motion. Thus

$$\frac{1}{\delta M}\frac{d}{dt}(\delta M) = \frac{1}{\rho}\frac{d}{dt}\rho + \frac{1}{\delta A}\frac{d}{dt}\delta A + \frac{1}{\delta z}\frac{d}{dt}\delta z = 0$$

But

$$\frac{1}{\delta A}\frac{d}{dt}\delta A = \frac{1}{\delta x}\frac{d}{dt}\delta x + \frac{1}{\delta y}\frac{d}{dt}\delta y$$

So that after changing the order of the differential operators, we obtain

$$\frac{1}{\rho}\frac{d\rho}{dt} + \frac{\delta u}{\delta x} + \frac{\delta v}{\delta y} + \frac{\delta w}{\delta z} = 0$$

which in the limit reduces to (4.2). Thus, it is apparent that the three-dimensional velocity divergence is approximately equal to the fractional rate of change of volume of a small element while the horizontal velocity divergence is approximately equal to the fraction rate of change of the horizontal area:

$$\frac{\partial u}{\partial x} + \frac{\partial v}{\partial y} \approx \frac{1}{\delta A}\frac{d}{dt}\delta A \tag{4.3}$$

4.2 Equation of Continuity in Isobaric Coordinates

We consider a fluid element of mass δM and cross-section area $\delta x\,\delta y$ which is confined between pressure surfaces p and $p - \delta p$ as shown in Fig. 4.2. Applying the hydrostatic approximation $\delta p = \rho g\,\delta z$ we may write

$$\delta M = \rho\,\delta x\,\delta y\,\delta z = \frac{\delta x\,\delta y\,\delta p}{g}$$

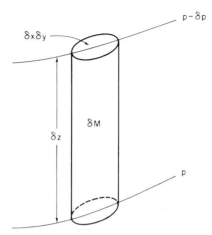

Fig. 4.2 An air column of fixed mass δM confined between two isobaric surfaces.

Since the mass of the fluid element is conserved following the motion,

$$\frac{1}{\delta M}\frac{d}{dt}\delta M = \frac{g}{\delta x\,\delta y\,\delta p}\frac{d}{dt}\left(\frac{\delta x\,\delta y\,\delta p}{g}\right) = 0$$

After differentiating using the chain rule and changing the order of the differential operators we obtain

$$\frac{1}{\delta x}\delta\left(\frac{dx}{dt}\right) + \frac{1}{\delta y}\delta\left(\frac{dy}{dt}\right) + \frac{1}{\delta p}\delta\left(\frac{dp}{dt}\right) = 0$$

or

$$\frac{\delta u}{\delta x} + \frac{\delta v}{\delta y} + \frac{\delta\omega}{\delta p} = 0$$

where we have defined $\omega \equiv dp/dt$. The field variable ω thus plays the same role in the isobaric coordinate system that w plays in the x, y, z system. Taking the limit $\delta x, \delta y, \delta p \to 0$ we obtain the continuity equation in the isobaric system

$$\left(\frac{\partial u}{\partial x} + \frac{\partial v}{\partial y}\right)_p + \frac{\partial\omega}{\partial p} = 0 \qquad (4.4)$$

This form of the continuity equation contains no reference to the density field, and does not involve time derivatives. The simplicity of (4.4) is one of the chief advantages of the isobaric coordinate system.

4.3 Vertical Motion

As previously mentioned, for synoptic scale motions the vertical velocity component is typically of the order of a few centimeters per second. Routine meteorological soundings, however, only give the wind speed to an accuracy of about a meter per second. Thus, in general the vertical velocity is not measured directly but must be inferred from the fields which are directly measured. One possible method of deducing the vertical velocity is based on integrating the continuity equation in the vertical. This method may be illustrated by considering the case of an incompressible fluid (the ocean, for example). For an incompressible fluid, $d\rho/dt = 0$ so that (4.2) becomes

$$\frac{\partial w}{\partial z} = -\left(\frac{\partial u}{\partial x} + \frac{\partial v}{\partial y}\right) \tag{4.5}$$

Integrating (4.5) with respect to z from the ground ($z = 0$) to a height h we obtain

$$w(h) - w(0) = -\int_0^h \left(\frac{\partial u}{\partial x} + \frac{\partial v}{\partial y}\right) dz = -h\left(\frac{\partial\langle u\rangle}{\partial x} + \frac{\partial\langle v\rangle}{\partial y}\right) \tag{4.6}$$

where the notation $\langle\ \rangle$ indicates the vertical average of a quantity. Thus for an incompressible fluid the difference between the vertical velocities at the top and bottom of a column is given by the depth of the column times the mean horizontal divergence.

For a compressible atmosphere, it is simplest to apply the isobaric co-ordinate form of the continuity equation. Integration of (4.4) with respect to pressure gives

$$\omega(p) = \omega(p_0) - \int_{p_0}^p \left(\frac{\partial u}{\partial x} + \frac{\partial v}{\partial y}\right)_p dp$$
$$= \omega(p_0) + (p_0 - p)\left(\frac{\partial\langle u\rangle}{\partial x} + \frac{\partial\langle v\rangle}{\partial y}\right)_p \tag{4.7}$$

which relates ω at any pressure p to $\omega(p_0)$ and the mean horizontal divergence in the column between the isobaric surfaces p_0 and p.

4.4 The Relationship between ω and w

Equation (4.7) can be converted into an expression involving the vertical velocity by expanding dp/dt in the x, y, z system and using scaling arguments to obtain an approximate relationship between ω and w. We have

$$\omega \equiv \frac{dp}{dt} = \frac{\partial p}{\partial t} + \mathbf{V}_h \cdot \nabla p + w\frac{\partial p}{\partial z} \tag{4.8}$$

Now, for synoptic scale motions, the horizontal velocity is geostrophic to a first approximation. Therefore we can write $\mathbf{V}_h = \mathbf{V}_g + \mathbf{V}'$, where $|\mathbf{V}'| \ll |\mathbf{V}_g|$. But $\mathbf{V}_g = (1/\rho f)\mathbf{k} \times \nabla p$ so that $\mathbf{V}_g \cdot \nabla p = 0$. Using this result plus the hydrostatic approximation, (4.8) may be rewritten as

$$\omega = \frac{\partial p}{\partial t} + \mathbf{V'} \cdot \nabla p - g\rho w \qquad (4.9)$$

Comparing the magnitudes of the three terms on the right in (4.9) we find that for synoptic scale motions

$$\frac{\partial p}{dt} \sim 10 \quad \text{mb day}^{-1}$$

$$\mathbf{V'} \cdot \nabla p \sim (1 \quad \text{m sec}^{-1})(0.01 \quad \text{mb km}^{-1}) \sim 1 \quad \text{mb day}^{-1}$$

$$g\rho w \sim 100 \quad \text{mb day}^{-1}$$

Thus, in the first approximation,

$$\omega = -\rho g w \qquad (4.10)$$

And (4.7) may be rewritten approximately as

$$w(p) = \frac{\rho_0 w(p_0)}{\rho} - \frac{1}{g\rho} \left(\frac{\partial \langle u \rangle}{\partial x} + \frac{\partial \langle v \rangle}{\partial y} \right)(p_0 - p) \qquad (4.11)$$

where ρ_0 is the density at pressure level p_0.

4.5 The Measurement of Horizontal Divergence

Application of (4.11) to infer the vertical velocity field requires knowledge of the horizontal divergence. In order to determine the horizontal divergence the partial derivatives $\partial u/\partial x$ and $\partial v/\partial y$ are generally estimated from the fields of u and v by using *finite difference* approximations. For example, to determine the horizontal divergence at the point x_0, y_0 in Fig. 4.3, we write

$$\frac{\partial u}{\partial x} + \frac{\partial v}{\partial y} \approx \frac{u(x_0 + d) - u(x_0 - d)}{2d} + \frac{v(y_0 + d) - v(y_0 - d)}{2d} \qquad (4.12)$$

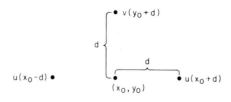

Fig. 4.3 Grid for estimation of the horizontal divergence.

However, for synoptic scale motions in midlatitudes the horizontal velocity is nearly in geostrophic equilibrium. Except for the small effect due to the variation of the Coriolis parameter (see Problem 1) the geostrophic wind is nondivergent, that is, $\partial u/\partial x$ and $\partial v/\partial y$ are nearly equal in magnitude but opposite in sign. Thus, the horizontal divergence is due primarily to the small departures of the wind from geostrophic balance. A ten percent error in evaluating one of the wind components in (4.12) can easily cause the estimated divergence to be in error by 100 percent. For this reason, the continuity equation method is not recommended for estimating the vertical motion field from observed horizontal winds. In Chapter 7 we will develop an alternative method, the so-called omega equation which is not nearly so sensitive to observational errors.

Problems

1. Show that the divergence of the geostrophic wind in isobaric coordinates is

$$\mathbf{V} \cdot \mathbf{V}_g = - \frac{\cot \phi}{fa} \frac{\partial \Phi}{\partial x}$$

   where a is the radius of the earth and ϕ is the latitude. What is the geostrophic divergence for a south wind of 10 m sec^{-1} at 43°N?

2. The following wind data were received from 50 km to the east, north, west, and south of a station respectively: 90°, 10 m sec^{-1}; 120°, 4 m sec^{-1}; 90°, 8 m sec^{-1}; 60°, 4 m sec^{-1}. Calculate the approximate horizontal divergence at the station.

3. Suppose that the wind speeds given in Problem 2 are each in error by $\pm 10\%$. What would be the percent error in the calculated horizontal divergence in the worst case?

4. The values of the horizontal divergence at various levels over a radiosonde station are measured as follows:

Pressure (mb)	$\mathbf{V} \cdot \mathbf{V}_h (\times 10^{-5} \text{ sec}^{-1})$
1000	+0.9
850	+0.6
700	+0.3
500	0
300	−0.6
100	−1.0

Compute the vertical velocity at each level assuming an isothermal atmosphere with temperature 260°K and letting $w = 0$ at 1000 mb.

5 | Circulation and Vorticity

In classical mechanics the principle of conservation of angular momentum is often invoked in the analysis of motions that involve rotation. This principle provides a powerful constraint on the behavior of rotating objects. Analogous conservation laws also apply to the rotational field of a fluid. However, it should be obvious that in a continuous medium such as the atmosphere the definition of "rotation" is more difficult than for a solid object.

Circulation and vorticity are the two primary measures of rotation in a fluid. Circulation, which is an integral quantity, is a *macroscopic* measure of rotation for a finite area of the fluid. Vorticity, however, is a vector field which gives a microscopic measure of the rotation at any point in the fluid.

5.1 The Circulation Theorem

By definition *circulation* is the line integral of the tangential component of the velocity around a closed path. If the path of integration is defined by the vector **l** we may write, as indicated in Fig. 5.1

$$C \equiv \oint_l \mathbf{V} \cdot d\mathbf{l} = \oint |\mathbf{V}| \cos \alpha \, dl$$

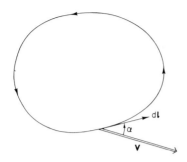

Fig. 5.1   Circulation about a closed loop l.

By convention the circulation is taken to be positive for a counterclockwise circuit about l.

That circulation is a measure of rotation can be seen from the following example. Suppose that a circular disk of fluid of radius r is in solid-body rotation at angular velocity Ω about the z axis. In this case $\mathbf{V} = \Omega \times \mathbf{r}$ where $\mathbf{r}$ is the distance to the axis of rotation. Thus the circulation about the edge of the disk is given by

$$C = \oint_l \mathbf{V} \cdot d\mathbf{l} = \int_0^{2\pi} \Omega r^2 \, d\lambda = 2\Omega \pi r^2$$

or

$$\frac{C}{\pi r^2} = 2\Omega$$

Hence, in the case of solid-body rotation the circulation divided by the area enclosed by the loop is just twice the angular velocity of rotation.

The circulation theorem is obtained by taking the line integral of Newton's second law for a closed chain of fluid particles. In the absolute coordinate system the result is

$$\oint \frac{d_a \mathbf{V}_a}{dt} \cdot d\mathbf{l} = -\oint \frac{\nabla p \cdot d\mathbf{l}}{\rho} - \oint \nabla \Phi \cdot d\mathbf{l} \tag{5.1}$$

The integrand on the left-hand side can be rewritten as

$$\frac{d_a \mathbf{V}_a}{dt} \cdot d\mathbf{l} = \frac{d}{dt}(\mathbf{V}_a \cdot d\mathbf{l}) - \mathbf{V}_a \cdot \frac{d}{dt}(d\mathbf{l})$$

or after observing that since $\mathbf{l}$ is a position vector, $d\mathbf{l}/dt \equiv \mathbf{V}_a$

$$\frac{d_a \mathbf{V}_a}{dt} \cdot d\mathbf{l} = \frac{d}{dt}(\mathbf{V}_a \cdot d\mathbf{l}) - \mathbf{V}_a \cdot d\mathbf{V}_a \tag{5.2}$$

Substituting (5.2) into (5.1) and using the fact that the line integral of a perfect differential is zero so that

$$\oint \nabla \Phi \cdot d\mathbf{l} = \oint d\Phi = 0$$

and

$$\oint \mathbf{V}_a \cdot d\mathbf{V}_a = \tfrac{1}{2} \oint d(\mathbf{V}_a \cdot \mathbf{V}_a) = 0$$

we obtain the circulation theorem:

$$\frac{dC_a}{dt} = \frac{d}{dt} \oint \mathbf{V}_a \cdot d\mathbf{l} = -\oint \frac{dp}{\rho} \tag{5.3}$$

The term on the right-hand side in (5.3) is called the *solenoidal* term. For a barotropic fluid, the density is a function only of pressure and the solenoidal term is zero. Thus, in a barotropic fluid the *absolute* circulation is conserved following the motion. This result, called Kelvin's circulation theorem, is a fluid analogue of angular momentum conservation in solid-body mechanics.

For meteorological analysis, it is more convenient to work with the relative circulation C rather than the absolute circulation since a portion of the absolute circulation, C_e, is due to the rotation of the earth about its axis. To compute C_e we apply Stokes theorem to the vector $\mathbf{V}_e$, where $\mathbf{V}_e = \mathbf{\Omega} \times \mathbf{r}$ is the velocity of the earth. Thus,

$$C_e = \oint \mathbf{V}_e \cdot d\mathbf{l} = \int_A \int (\nabla \times \mathbf{V}_e) \cdot \mathbf{n} \, dA$$

But $(\nabla \times \mathbf{V}_e) \cdot \mathbf{n} = 2\Omega \sin \phi \equiv f$ is just the Coriolis parameter. Hence the circulation due to the rotation of the earth is $C_e = 2\Omega \sin \bar{\phi} A$, where $\bar{\phi}$ denotes the average value of latitude over the area element A. Finally, we may write

$$C = C_a - C_e = C_a - 2\Omega F \tag{5.4}$$

where $F = A \sin \bar{\phi}$ is the projection of the area element A on the equatorial plane as shown in Fig. 5.2. Differentiating (5.4) following the motion, and

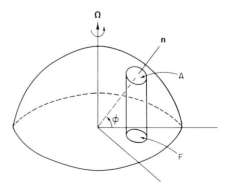

Fig. 5.2 Area subtended on the equatorial plane by a closed loop at latitude ϕ.

substituting from (5.3) we obtain the Bjerknes circulation theorem:

$$\frac{dC}{dt} = - \oint \frac{dp}{\rho} - 2\Omega \frac{dF}{dt} \qquad (5.5)$$

For a barotropic fluid, (5.5) can be integrated following the motion from an initial state designated by subscript 1 to a final state designated by subscript 2, yielding

$$C_2 - C_1 = -2\Omega(A_2 \sin \phi_2 - A_1 \sin \phi_1) \qquad (5.6)$$

Equation (5.6) indicates that in a barotropic fluid the relative circulation for a closed chain of fluid particles will be changed if either the horizontal area enclosed by the loop changes or the average latitude changes.

Example: Suppose that the air within a circular region of radius 100 km centered at the equator is initially motionless with respect to the earth. If this circular air mass were moved to the North Pole along an isobaric surface, the circulation about the circumference would be

$$C = -2\Omega\pi r^2 (\sin \pi/2 - \sin 0)$$

Thus the mean tangential velocity at the radius $r = 100$ km would be

$$\overline{V} = \frac{C}{2\pi r} = -\Omega r \approx -7 \quad \text{m sec}^{-1}$$

The negative sign here indicates that the air has acquired anticyclonic relative circulation.

We have yet to consider the role of the solenoidal term in the circulation theorem. The generation of circulation by pressure–density solenoids can be effectively illustrated by considering the development of a seabreeze circulation. The situation is shown in Fig. 5.3. The mean temperature in the air over the ocean is colder than the mean temperature over the adjoining land. Thus if the pressure is uniform at ground level, the isobaric surfaces above the ground will slope downward toward the ocean while the isosteric surfaces (isolines of specific volume) slope downward toward the land. To compute the acceleration as a result of the intersection of the pressure–density surfaces we apply the circulation theorem on a circuit in a vertical plane perpendicular to the coastline. Substituting the ideal gas law into (5.3) we obtain

$$\frac{dC_a}{dt} = - \oint RT \, d \ln p$$

Evaluating the line integral for the circuit shown in Fig. 5.3 we see that there is a contribution only for the vertical segments of the loop since the horizontal segments are taken at constant pressure (the slope of the isobaric surface for

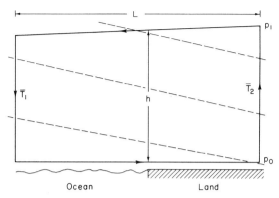

Fig. 5.3 Application of the circulation theorem to the seabreeze problem. The closed heavy solid line is the loop about which the circulation is to be evaluated. Dashed lines indicate isosteric surfaces.

the upper segment is neglected here since it is small compared to the slopes of the isosteric surfaces). The resulting rate of increase in the circulation is

$$\frac{dC_a}{dt} = R \ln \left(\frac{p_0}{p_1}\right) (\overline{T}_2 - \overline{T}_1) > 0$$

Letting $\bar{v}$ be the mean tangential velocity along the circuit, we find that

$$\frac{d\bar{v}}{dt} = \frac{R \ln (p_0/p_1)}{2(h + L)} (\overline{T}_2 - \overline{T}_1) \tag{5.7}$$

As a specific example we let $p_0 = 1000$ mb, $p_1 = 900$ mb, $\overline{T}_2 - \overline{T}_1 = 10°C$, $L = 20$ km, and $h = 1$ km. The acceleration is then

$$\frac{d\bar{v}}{dt} \approx 0.685 \quad \text{cm sec}^{-2}$$

so that in the absence of frictional retarding forces the wind would reach a speed of 20 m sec^{-1} in about one hour. In reality, however, frictional drag (which is roughly proportional to the square of the wind speed) would quickly retard the acceleration and eventually a balance would be reached between the generation of kinetic energy by the pressure–density solenoids and dissipation by friction.

5.2 Vorticity

Vorticity, the microscopic measure of rotation in a fluid, is a vector field defined as the curl of velocity. In dynamic meteorology we are primarily concerned with the vertical component of vorticity $\zeta \equiv \mathbf{k} \cdot \mathbf{V} \times \mathbf{V}$. In cartesian coordinates the vertical component is just

$$\zeta = \frac{\partial v}{\partial x} - \frac{\partial u}{\partial y}$$

Alternatively, the vertical component of vorticity may be defined as the circulation about a loop in the horizontal plane divided by the area enclosed, in the limit where the area approaches zero:

$$\zeta \equiv \lim_{A \to 0} \frac{\oint \mathbf{V} \cdot d\mathbf{l}}{A} \tag{5.8}$$

This latter definition makes explicit the relationship between circulation and vorticity discussed in the introduction to this chapter. The equivalence of these two definitions of ζ may be easily shown by considering the circulation about a rectangular element of area in the x, y plane as shown in Fig. 5.4.

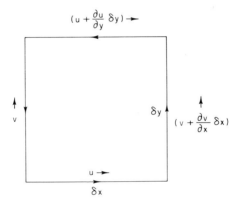

Fig. 5.4   The relationship between circulation and vorticity for an area element in the horizontal plane.

With the aid of the figure we see that the circulation is

$$C = u\delta x + \left(v + \frac{\partial v}{\partial x}\delta x\right)\delta y - \left(u + \frac{\partial u}{\partial y}\delta y\right)\delta x - v\,\delta y$$

$$= \left(\frac{\partial v}{\partial x} - \frac{\partial u}{\partial y}\right)\delta x\,\delta y$$

Dividing through by the area $\delta A = \delta x\,\delta y$, we get

$$\frac{C}{\delta A} = \frac{\partial v}{\partial x} - \frac{\partial u}{\partial y} \equiv \zeta$$

In more general terms the relationship between vorticity and circulation is given simply by Stokes's theorem applied to the velocity vector:

$$\oint \mathbf{V} \cdot d\mathbf{l} = \int_\sigma \int (\nabla \times \mathbf{V}) \cdot \mathbf{n} \, d\sigma$$

Here σ is the area enclosed by the curve $\mathbf{l}$, and $\mathbf{n}$ is a unit normal to the area element $d\sigma$. Thus, Stokes's theorem states that the circulation about any closed loop is equal to the integral of the normal component of vorticity over the area enclosed by the loop. Hence, for a finite area, the circulation divided by the area gives the *average* normal component of vorticity in the region. As a consequence, the vorticity of a fluid in solid-body rotation is just twice the angular velocity of rotation. Vorticity may thus be regarded as a measure of the local angular velocity of the fluid.

Physical interpretation of vorticity may be facilitated by considering the vertical component of vorticity in the natural coordinate system (see Section 3.2.1). If we compute the circulation about the infinitesimal loop shown in Fig. 5.5 we obtain

$$C = V[\delta s + d(\delta s)] - \left(V + \frac{\partial V}{\partial n} \delta n \right) \delta s$$

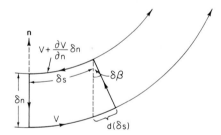

Fig. 5.5 Circulation for an infinitesimal loop in the natural coordinate system.

But from Fig. 5.5 we see that

$$d(\delta s) = \delta \beta \, \delta n$$

where $\delta \beta$ is the angular change in the wind direction in the distance δs. Hence

$$C = \left(-\frac{\partial V}{\partial n} + V \frac{\delta \beta}{\delta s} \right) \delta n \, \delta s$$

or in the limit $\delta n, \delta s \to 0$

$$\zeta = \lim_{\delta n, \delta s \to 0} \frac{C}{(\delta n \, \delta s)} = -\frac{\partial V}{\partial n} + \frac{V}{R_s} \tag{5.9}$$

where R_s is the radius of curvature of the streamlines [Equation (3.20)]. It is now apparent that the net vertical vorticity component is the result of the sum of two parts: (1) the rate of change of wind speed normal to the direction of flow, $-\partial V/\partial n$, called the *shear vorticity*; and (2) the turning of the wind along a streamline, V/R_s called the *curvature vorticity*. Thus, even straight-line motion may have vorticity if the speed changes normal to the flow axis. For example, in the jet stream shown schematically in Fig. 5.6a there will be

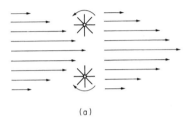

(a)

Fig. 5.6 Two types of two-dimensional flow: (a) linear shear flow with vorticity and (b) curved flow with zero vorticity.

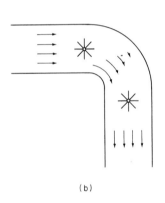

(b)

cyclonic vorticity north of the velocity maximum and anticyclonic vorticity to the south as is easily recognized when the motion of a small exploring paddle wheel is considered. The lower paddle wheel shown in the figure will be turned in a clockwise direction since the wind force on its northern blades is stronger than the force on the blades to the south of the axis of the wheel.

On the other hand, curved flow may have zero vorticity provided that the shear vorticity is equal and opposite to the curvature vorticity. This will be the case in the example shown in Fig. 5.6b where a frictionless fluid with zero relative vorticity upstream flows around a bend in a canal. The water along the inner margin of the channel flows faster in just the right proportion so that the paddle wheel does not turn.

5.3 Potential Vorticity

Since potential temperature θ is conserved following the motion in adiabatic flow, a parcel of air which moves adiabatically will remain on the same potential temperature surface. Now from the definition of potential temperature it can be shown that for constant θ

$$\rho \propto p^{c_v/c_p}$$

Therefore, density is a function of pressure alone on an adiabatic surface and the solenoidal term in the circulation theorem will vanish:

$$\oint \frac{dp}{\rho} \propto \oint dp^{(1-c_v/c_p)} = 0$$

Thus, for adiabatic motion the circulation theorem evaluated on a constant-θ surface reduces to the same form as in a barotropic fluid, that is,

$$\frac{d}{dt}(C + 2\Omega A \sin \phi) = 0 \tag{5.10}$$

where C is evaluated for a closed loop encompassing the area A on an isentropic surface. If we assume that the isentropic surface is approximately horizontal, and recall that the vertical relative vorticity component is given by

$$\zeta = \lim_{A \to 0} \frac{C}{A}$$

we can write the integral of (5.10) for an infinitesimal parcel of air as

$$A(\zeta + f) = \text{const} \tag{5.11}$$

where $f = 2\Omega \sin \phi$ is the Coriolis parameter. Now suppose that the parcel is confined between potential temperature surfaces θ_0 and $\theta_0 + \delta\theta$ which are separated by a distance δp as shown in Fig. 5.7. The mass of the parcel, $M = A\delta p/g$, must be conserved following the motion. Therefore

$$A = \frac{\text{const}}{\delta p} = \text{const} \times \frac{\delta\theta}{\delta p}$$

since $\delta\theta$ is a constant. Substituting into (5.11) to eliminate A and taking the limit $\delta p \to 0$, we obtain

$$(\zeta + f)\frac{\partial\theta}{\partial p} = \text{const} \tag{5.12}$$

Equation (5.12) expresses the conservation of potential *vorticity* in adiabatic, frictionless motion. The term potential vorticity is used, as we shall see later,

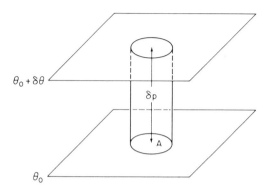

Fig. 5.7   An air column moving isentropically.

in connection with several other slightly different mathematical expressions. In essence, however, the potential vorticity is always in some sense a measure of the ratio of the absolute vorticity to the effective depth of the vortex. In (5.12), for example, the effective depth is just the distance between potential temperature surfaces measured in pressure units.

In a homogeneous incompressible fluid the theorem of potential vorticity conservation takes a somewhat simpler form. In this case since density is a constant we have

$$A = \frac{\text{const}}{\delta z}$$

where δz is the height of the parcel. Again substituting to eliminate A in (5.11) we get

$$\frac{\zeta + f}{\delta z} = \text{const} \tag{5.13}$$

The conservation of potential vorticity is a powerful constraint on the large-scale motions of the atmosphere. This can be illustrated by considering the flow of air over a large mountain barrier. The situation for westerly flow is shown in Fig. 5.8. In Fig. 5.8a a vertical cross section of the flow is shown. We suppose that upstream of the mountain barrier the flow is a uniform zonal flow so that $\zeta = 0$. If the flow is adiabatic, each column of air confined between the potential temperature surfaces θ_0 and $\theta_0 + \delta\theta$ remains between those surfaces as it crosses the mountain. For this reason, a potential temperature surface near the ground must approximately follow the ground contours. However, ordinarily, a potential temperature surface several kilometers above the ground will not be deflected much in the vertical by the ground contours. Thus, as an air column begins to cross the mountain barrier its vertical

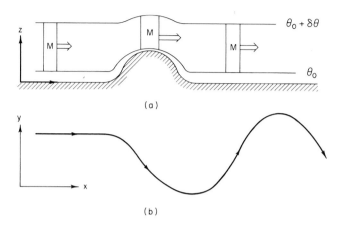

Fig. 5.8   Westerly flow over a topographic barrier: (a) the depth of a column as a function of x, (b) the trajectory of a parcel in the x, y plane.

extent decreases, hence as is obvious from (5.12) in order to conserve potential vorticity the relative vorticity must become negative. Thus the air will acquire anticyclonic vorticity and move southward as shown in the x, y plane profile in Fig. 5.8b. Now when the air parcel has passed over the mountain and the column has returned to its original depth it will be south of its original latitude so that f will be smaller and the relative vorticity must now be positive—that is, the trajectory must have cyclonic curvature. When the parcel returns to its original latitude, it will still have a northward velocity component and will continue northward gradually acquiring anticyclonic curvature until its meridional motion direction is reversed. Thus, the parcel will move downstream conserving potential vorticity by following a wavelike trajectory in the horizontal plane. Therefore, steady westerly flow over a mountain barrier will result in a cyclonic flow pattern immediately to the east of the barrier (the lee side trough) followed by an alternating series of ridges and troughs downstream.

The situation for easterly flow impinging on a mountain barrier is quite different. Let us assume that the motion were to remain uniform until the flow reached the mountain. In that case as an air column began to move over the barrier from the east it would have to acquire anticyclonic curvature in order to conserve potential vorticity. Thus, the parcel trajectory would begin to curve northward toward higher latitudes, and the resulting increase in f would force the relative vorticity to become even more strongly negative to conserve potential vorticity. As a result the parcel would eventually curve back toward the east and no steady-state potential vorticity conserving flow pattern would be possible.

For this reason, easterly flow, unlike westerly flow, must feel the influence of the mountain barrier upstream from the barrier. As indicated schematically in Fig. 5.9, a westward-moving air column begins to curve cyclonically before

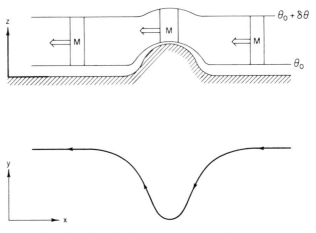

Fig. 5.9   Easterly flow over a topographic barrier.

it reaches the mountain. The resulting positive relative vorticity is balanced by a decrease in f so that potential vorticity is conserved. As the column moves to the top of the mountain, it continues to move equatorward so that the decrease in depth is balanced by a decrease in f. Finally, as the parcel moves down the mountain toward the west, the process is simply reversed with the result that some distance downstream from the mountain the air column again is moving westward at its original latitude. Thus, the dependence of the Coriolis parameter on latitude creates a dramatic difference between westerly and easterly flow over a topographic barrier. In the case of a westerly wind the barrier generates a wavelike disturbance in the streamlines which propagates downstream. But in the case of an easterly wind, the disturbance in the streamlines damps out away from the barrier.

Equation (5.13) also indicates that in a barotropic fluid a change in the depth is dynamically analogous to a change in the Coriolis parameter. This effect can easily be demonstrated in a rotating cylindrical vessel filled with an incompressible fluid. For such a fluid in solid-body rotation, the equilibrium shape of the free surface, determined by a balance between the radial pressure gradient and centrifugal forces, is parabolic. Thus, as shown in Fig. 5.10, if a column of fluid moves radially outward it must stretch vertically. According to (5.13) the relative vorticity must then increase to keep the product $(\zeta + f)/\delta z$ constant. The same result would apply if a column of fluid on a rotating sphere were moved equatorward without change in depth. In this case ζ would have to increase to offset the decrease of f. Therefore, in a barotropic

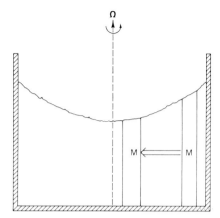

Fig. 5.10 The dependence of depth on radius in a rotating cylindrical vessel.

fluid a decrease of depth with increasing latitude has the same effect on the relative vorticity as the increase of the Coriolis force with latitude.

5.4 The Vorticity Equation

In the previous section we have discussed the time evolution of the vertical component of vorticity for the special case of adiabatic frictionless flow. In the present section we will use the equations of motion to derive an equation for the time rate of change of vorticity without limiting the validity to adiabatic motion.

5.4.1 CARTESIAN COORDINATE FORM

For motions of synoptic scale, the vorticity equation can be derived using the approximate horizontal momentum equations (2.24) and (2.25). We differentiate the x component equation with respect to y and the y component equation with respect to x:

$$\frac{\partial}{\partial y}\left(\frac{\partial u}{\partial t} + u\frac{\partial u}{\partial x} + v\frac{\partial u}{\partial y} + w\frac{\partial u}{\partial z} - fv = -\frac{1}{\rho}\frac{\partial p}{\partial x}\right) \qquad (5.14)$$

$$\frac{\partial}{\partial x}\left(\frac{\partial v}{\partial t} + u\frac{\partial v}{\partial x} + v\frac{\partial v}{\partial y} + w\frac{\partial v}{\partial z} + fu = -\frac{1}{\rho}\frac{\partial p}{\partial y}\right) \qquad (5.15)$$

Subtracting (5.14) from (5.15) and recalling that $\zeta = \partial v/\partial x - \partial u/\partial y$ we obtain the vorticity equation

$$\frac{\partial \zeta}{\partial t} + u\frac{\partial \zeta}{\partial x} + v\frac{\partial \zeta}{\partial y} + w\frac{\partial \zeta}{\partial z} + (\zeta + f)\left(\frac{\partial u}{\partial x} + \frac{\partial v}{\partial y}\right)$$
$$+ \left(\frac{\partial w}{\partial x}\frac{\partial v}{\partial z} - \frac{\partial w}{\partial y}\frac{\partial u}{\partial z}\right) + v\frac{df}{dy} = \frac{1}{\rho^2}\left[\frac{\partial \rho}{\partial x}\frac{\partial p}{\partial y} - \frac{\partial \rho}{\partial y}\frac{\partial p}{\partial x}\right] \qquad (5.16)$$

Using the fact that the Coriolis parameter depends only on y so that $df/dt = v(df/dy)$, (5.16) may be rewritten in the form

$$\frac{d}{dt}(\zeta + f) = -(\zeta + f)\left(\frac{\partial u}{\partial x} + \frac{\partial v}{\partial y}\right) - \left(\frac{\partial w}{\partial x}\frac{\partial v}{\partial z} - \frac{\partial w}{\partial y}\frac{\partial u}{\partial z}\right) + \frac{1}{\rho^2}\left(\frac{\partial \rho}{\partial x}\frac{\partial p}{\partial y} - \frac{\partial \rho}{\partial y}\frac{\partial p}{\partial x}\right)$$

$$(5.17)$$

Equation (5.17) states that the rate of change of the absolute vorticity follow-ing the motion is given by the sum of the three terms on the right, called the divergence term, the tilting or twisting term, and the solenoidal term, respec-tively.

Generation of vorticity by horizontal divergence, the first term on the right in (5.17), is the fluid analogue of the change in angular velocity resulting from a change in the moment of inertia of a solid body when angular momentum is conserved. If there is positive horizontal divergence, the area enclosed by a chain of fluid parcels will increase with time; and if circulation is to be conserved the average absolute vorticity of the enclosed fluid must decrease. This mechanism for changing vorticity is very important in synoptic scale disturbances.

The second term on the right in (5.17) represents vertical vorticity which is generated by the "tilting" of horizontally oriented components of vorticity into the vertical by a nonuniform vertical motion field. This mechanism is illustrated in Fig. 5.11. As shown in the figure we consider a region where the y component of velocity is increasing with height so that there is a component of shear vorticity oriented in the negative x direction. This is indicated by the

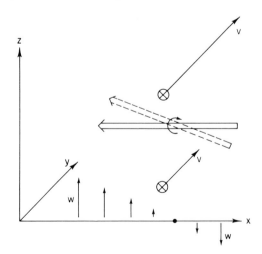

Fig. 5.11 Vorticity generation by the tilting of a horizontal vortex tube (double arrow).

double arrow in Fig. 5.11. If at the same time there is a vertical motion field in which w decreases with increasing x, the vertical motion will tend to tilt the vortex tube initially oriented parallel to x so that it has a component in the vertical. Thus, if $\partial v/\partial z > 0$ and $\partial w/\partial x < 0$, there will be a generation of positive vertical vorticity.

Finally, the third term on the right in (5.17) is just the microscopic equivalent of the solenoidal term in the circulation theorem (5.5). To show this equivalence we may apply Stokes's theorem to the solenoidal term to get

$$\oint -\alpha\, dp \equiv -\oint \alpha\nabla p \cdot d\mathbf{l} = -\int_\sigma \int \nabla \times (\alpha\nabla p) \cdot \mathbf{k}\, d\sigma$$

where σ is the horizontal area bounded by the curve $\mathbf{l}$. Applying the vector identity $\nabla \times (\alpha\nabla p) \equiv \nabla\alpha \times \nabla p$ we see that

$$\oint -\alpha\, dp = -\int_\sigma \int (\nabla\alpha \times \nabla p) \cdot \mathbf{k}\, d\sigma$$

But the solenoidal term in the vorticity equation can be written

$$-\left(\frac{\partial\alpha}{\partial x}\frac{\partial p}{\partial y} - \frac{\partial\alpha}{\partial y}\frac{\partial p}{\partial x}\right) = -\mathbf{k} \cdot (\nabla\alpha \times \nabla p)$$

Comparing the right-hand sides of these two expressions, we see that the solenoidal term in the vorticity equation is just the limit of the solenoidal term in the circulation theorem divided by the area when the area goes to zero.

5.4.2 THE VECTOR VORTICITY EQUATION

In the previous subsection we have used the simplified cartesian form of the equations of motion to derive the vertical component vorticity equation. To obtain a more general form for the vertical vorticity equation it is convenient to begin with (3.9), the vector horizontal momentum equation:

$$\frac{\partial\mathbf{V}_h}{dt} + (\mathbf{V}_h \cdot \nabla_h)\mathbf{V}_h + w\frac{\partial\mathbf{V}_h}{\partial z} = -\frac{1}{\rho}\nabla_h p - 2\mathbf{\Omega} \times \mathbf{V}_h \qquad (5.18)$$

The vector vorticity equation may then be formally obtained by merely taking the vertical component of the curl of (5.18). However, to facilitate this operation it is desirable to first apply the vector identity

$$(\mathbf{V}_h \cdot \nabla_h)\mathbf{V}_h \equiv \nabla_h\left(\frac{\mathbf{V}_h \cdot \mathbf{V}_h}{2}\right) + \mathbf{k} \times \mathbf{V}_h\zeta \qquad (5.19)$$

where $\zeta = \mathbf{k} \cdot (\nabla \times \mathbf{V})$ and rewrite (5.18) as

$$\frac{\partial\mathbf{V}_h}{\partial t} = -\nabla_h\left(\frac{\mathbf{V}_h \cdot \mathbf{V}_h}{2}\right) - \alpha\nabla p - \mathbf{k} \times \mathbf{V}_h\zeta - w\frac{\partial\mathbf{V}_h}{\partial z} - 2\mathbf{\Omega} \times \mathbf{V}_h \qquad (5.20)$$

Then we apply the vector operator $\mathbf{k} \cdot \nabla \times$ to (5.20) and after some rearrangement of terms obtain

$$\frac{\partial \zeta}{\partial t} = -\mathbf{V}_h \cdot \nabla_h (\zeta + f) - w \frac{\partial \zeta}{\partial z} - (\zeta + f)\nabla_h \cdot \mathbf{V}_h$$

$$+ \mathbf{k} \cdot \left(\frac{\partial \mathbf{V}_h}{\partial z} \times \nabla w\right) - \mathbf{k} \cdot (\nabla \alpha \times \nabla p) \tag{5.21}$$

where we have used the facts that for any scalar ϕ, $\nabla \times \nabla \phi = 0$, and for any vectors $\mathbf{a}$ and $\mathbf{b}$

$$\nabla \times (\mathbf{a} \times \mathbf{b}) = (\nabla \cdot \mathbf{b})\mathbf{a} - (\mathbf{a} \cdot \nabla)\mathbf{b}$$

5.5 Scale Analysis of the Vorticity Equation

In Section 2.4 we showed that the equations of motion could be simplified for synoptic scale motions by evaluating the order of magnitude of various terms. The same technique can also be applied to the vorticity equation. To scale the vorticity equation we choose characteristic scales for the field variables based on typical observed magnitudes for synoptic-scale motions as follows:

$$U \sim 10^3 \text{ cm sec}^{-1}$$ Horizontal velocity
$$W \sim 1 \text{ cm sec}^{-1}$$ Vertical velocity
$$L \sim 10^8 \text{ cm}$$ Length scale
$$H \sim 10^6 \text{ cm}$$ Depth scale
$$\delta p \sim 10^4 \text{ dynes cm}^{-2}$$ Horizontal pressure fluctuations
$$\delta \rho / \rho \sim 0.02$$ Fractional density fluctuation
$$L/U \sim 10^5 \text{ sec}$$ Time scale
$$f_0 \sim 10^{-4} \text{ sec}^{-1}$$ Coriolis parameter

$$\frac{df}{dy} \equiv \beta \sim 10^{-13} \text{ cm sec}^{-1}$$ " Beta " parameter

Again we have chosen an advective time scale because the vorticity pattern, like the pressure pattern, tends to move at a speed comparable to the horizontal wind speed. Using these scales to evaluate the magnitudes of the terms in (5.16) we note first that

$$\zeta = \frac{\partial v}{\partial x} - \frac{\partial u}{\partial y} \lesssim \frac{U}{L} \sim 10^{-5} \quad \text{sec}^{-1}$$

where $\lesssim$ means less than or equal to in order of magnitude. Comparing with the Coriolis parameter we see that $\zeta/f_0 \lesssim 10^{-1}$ so that for midlatitude synoptic systems the relative vorticity is small compared to the earth's vorticity. Hence, ζ may be neglected compared to f in the divergence term:

$$(\zeta + f)\left(\frac{\partial u}{\partial x} + \frac{\partial v}{\partial y}\right) \approx f\left(\frac{\partial u}{\partial x} + \frac{\partial v}{\partial y}\right)$$

The magnitudes of the various terms in (5.16) are as follows

$$\frac{\partial \zeta}{\partial t}, u\frac{\partial \zeta}{\partial x}, v\frac{\partial \zeta}{\partial y} \sim \frac{U^2}{L^2} \sim 10^{-10} \quad \sec^{-2}$$

$$w\frac{\partial \zeta}{\partial z} \sim \frac{UW}{LH} \sim 10^{-11} \quad \sec^{-2}$$

$$v\frac{df}{dy} \sim U\beta \sim 10^{-10} \quad \sec^{-2}$$

$$f\left(\frac{\partial u}{\partial x} + \frac{\partial v}{\partial y}\right) \lesssim \frac{f_0 U}{L} \sim 10^{-9} \quad \sec^{-2}$$

$$\left(\frac{\partial w}{\partial x}\frac{\partial v}{\partial z} - \frac{\partial w}{\partial y}\frac{\partial u}{\partial z}\right) \lesssim \frac{WU}{HL} \sim 10^{-11} \quad \sec^{-2}$$

$$\frac{1}{\rho^2}\left(\frac{\partial \rho}{\partial x}\frac{\partial p}{\partial y} - \frac{\partial \rho}{\partial y}\frac{\partial p}{\partial x}\right) \lesssim \frac{\delta\rho\,\delta p}{\rho^2 L^2} \sim 2 \times 10^{-11} \quad \sec^{-2}$$

The symbol $\lesssim$ is used for the last three terms because in each case it is possible that the two parts of the expression might partially cancel so that the actual magnitude is less than indicated. In fact, comparing the magnitudes of the various terms we see that this must be true for the divergence term, because if $\partial u/\partial x$ and $\partial v/\partial y$ were not nearly equal and opposite, the divergence term would be an order of magnitude greater than any other term and the equation could not be satisfied. Therefore, scale analysis of the vorticity equation indicates that synoptic scale motions must be quasinondivergent. In order that the divergence term be small enough to be balanced by the vorticity advection terms we see that

$$\left(\frac{\partial u}{\partial x} + \frac{\partial v}{\partial y}\right) \lesssim 10^{-6} \quad \sec^{-1}$$

so that the horizontal divergence must be small compared to the vorticity in synoptic scale systems.

Retaining now only the terms of order 10^{-10} $\sec^{-2}$ in the vorticity equation, we obtain as the first approximation for synoptic scale motions

$$\frac{d_h(\zeta + f)}{dt} = -f\left(\frac{\partial u}{\partial x} + \frac{\partial v}{\partial y}\right) \tag{5.22}$$

where

$$\frac{d_h}{dt} = \frac{\partial}{\partial t} + u\frac{\partial}{\partial x} + v\frac{\partial}{\partial y}$$

Equation (5.22) states that as a first approximation the change of absolute vorticity following the motion on the synoptic scale is entirely due to the divergence effect. This approximation does not remain valid, however, in the vicinity of atmospheric fronts. The horizontal scale of variation in frontal zones is only ~ 100 km, and for this scale the vertical advection, tilting, and solenoidal terms all may become as large as the divergence term.

For the case of an incompressible barotropic fluid, the potential vorticity theorem can be derived from the approximate vorticity equation quite easily. Now for an incompressible fluid the continuity equation takes the form

$$\frac{\partial u}{\partial x} + \frac{\partial v}{\partial y} = -\frac{\partial w}{\partial z}$$

so that the vorticity equation (5.22) may be written

$$\frac{d_h(\zeta + f)}{dt} = (\zeta + f)\frac{\partial w}{\partial z} \qquad (5.23)$$

(we have here retained the relative vorticity in the divergence term for reasons which will become apparent). We showed in Section 3.4 that in a barotropic fluid the thermal wind vanishes so that the geostrophic wind is independent of height. Letting the vorticity in (5.23) be approximated by the vorticity of the geostrophic wind ($\zeta = \zeta_g$), we can integrate vertically from $z = D_1$ to $z = D_2$ to get

$$(D_2 - D_1)\frac{d_h(\zeta_g + f)}{dt} = (\zeta_g + f)[w(D_2) - w(D_1)]$$

But since $w \equiv dz\,dt$ we can write

$$w(D_2) = \frac{dD_2}{dt} \qquad \text{and} \qquad w(D_1) = \frac{dD_1}{dt}$$

so that letting $H = D_2 - D_1$ be the depth of the fluid column we get

$$\frac{1}{(\zeta_g + f)}\frac{d_h(\zeta_g + f)}{dt} = \frac{1}{H}\frac{d_h(H)}{dt}$$

or

$$\frac{d_h}{dt}\left(\frac{\zeta_g + f}{H}\right) = 0$$

which is just the potential vorticity theorem for a barotropic fluid.

Problems

1. What is the circulation about a square of 1000 km on a side for an easterly (that is, westward flowing) current which decreases in magnitude toward the north at a rate of 10 m sec^{-1}/500 km? What is the mean relative vorticity in the square?

2. A cylindrical column of air at 30°N with radius 100 km expands to twice its original radius. If the air is initially at rest, what is the mean tangential velocity at the perimeter after expansion?

3. An air parcel at 30°N moves northward conserving absolute vorticity. If its initial relative vorticity is 5×10^{-5} sec^{-1}, what is its relative vorticity upon reaching 90°N?

4. An air column at 60°N with $\zeta = 0$ initially stretches from the surface to a fixed tropopause at 10 km height. If the air column moves until it is over a mountain barrier 2.5-km high at 45°N, what is its absolute vorticity and relative vorticity as it passes the mountain top?

5. Find the average vorticity within a cylindrical annulus of inner radius 200 km and outer radius 400 km if the tangential velocity distribution is given by $V = 10^6/r$ m sec^{-1} where r is in meters. What is the average vorticity within the inner circle of radius 200 km?

6. Compute the rate of change of circulation about a square in the x, y plane with sides of 1000-km length if temperature increases eastward at a rate of 1°C/200 km and pressure increases northward at a rate of 1 mb/200 km. The pressure at the origin is 1000 mb.

7. Verify the identity (5.19) by expanding the vectors in cartesian components.

8. Derive a formula for the dependence of depth on radius for an incompressible fluid in solid-body rotation in a cylindrical tank. Let H be the depth at the center of the tank, Ω the angular velocity of the tank, and a the radius of the tank.

9. By how much does the relative vorticity change for a column of fluid in a rotating cylinder if the column is moved from the center of the tank to a distance 50 cm from the center? The tank is rotating at the rate of 20 revolutions per minute, the depth of the fluid at the center is 10 cm and the radius of the tank is 1 m.

Suggested References

Batchelor, *An Introduction to Fluid Dynamics,* has an extensive treatment of the general applications of the concepts of vorticity and circulation in fluid dynamics. Of special interest is the section on motion in a thin layer on a rotating sphere.

Milne-Thompson, *Theoretical Hydrodynamics,* is a classical text which discusses vortex motion for a number of idealized fluid models. Vector methods and notation are heavily used.

Chapter

6 | The Planetary Boundary Layer

In deriving the idealized flow fields discussed in Chapter 3 we assumed in all cases that frictional forces were sufficiently small that they could be neglected. For synoptic scale motions in the free atmosphere (that is, that part of the atmosphere far enough removed from the ground to be free of direct surface friction effects), this approximation is probably quite valid. However, in the lowest kilometer of the atmosphere the vertical viscous force generally is comparable in magnitude to the pressure gradient and Coriolis forces. This region of the atmosphere is referred to as the *planetary boundary layer*. It contains nearly ten percent of the mass of the atmosphere.

Because the planetary boundary layer is a turbulent boundary layer, a rigorous mathematical theory for the structure of the velocity field in the layer is not possible. In order to satisfactorily study the structure of the planetary boundary layer we would require detailed knowledge of the structure and amplitude of the turbulent eddies which are responsible for the vertical momentum transport. Even a cursory examination of this subject is outside the scope of this text. In the present chapter it will be assumed that the frictional force in turbulent flow may be represented in the same manner as done in laminar flow by introducing an *eddy viscosity* coefficient.

6.1   The Mixing Length Theory

The concept of an eddy viscosity coefficient was mentioned briefly in Section 1.2.3. In this section we will discuss in more detail the heuristic argument of the famous fluid dynamicist L. Prandtl which provides some theoretical basis for estimating the magnitude of the eddy viscosity. Prandtl's basic idea was that the momentum transport of the small-scale eddy motions may be parameterized in terms of the large-scale mean flow.

In order to understand the basis of this parameterization we must first consider the derivation of mean flow equations for a turbulent fluid. (In fact we have been dealing with such equations throughout this book; however, we have not previously explicitly considered the partitioning of the flow between turbulent eddies and mean fields.) In a turbulent fluid the velocity measured at a point generally fluctuates rapidly in time as eddies of various scales pass the point. In order that our velocity measurements be truly representative of the large-scale flow it is thus necessary to average the flow over an interval of time long enough to average out the eddy fluctuations but still short enough to preserve the trends in the large-scale flow field. To do this we use angle brackets to define the mean velocity $\langle \mathbf{V} \rangle$ as the average in time at a given point. The instantaneous velocity is then

$$\mathbf{V} = \langle \mathbf{V} \rangle + \mathbf{V}'$$

where $\mathbf{V}'$ designates the deviation from the mean at any time. $\mathbf{V}'$ is thus associated with the turbulent eddies.

We now apply this averaging scheme to the horizontal equations of motion. We have previously written these equations in the approximate form:

$$\frac{\partial u}{\partial t} + u\frac{\partial u}{\partial x} + v\frac{\partial u}{\partial y} + w\frac{\partial u}{\partial z} - fv = -\frac{1}{\rho}\frac{\partial p}{\partial x} \tag{6.1}$$

$$\frac{\partial v}{\partial t} + u\frac{\partial v}{\partial x} + v\frac{\partial v}{\partial y} + w\frac{\partial v}{\partial z} + fu = -\frac{1}{\rho}\frac{\partial p}{\partial y} \tag{6.2}$$

Using the continuity equation,

$$\frac{\partial \rho}{\partial t} + \frac{\partial}{\partial x}(\rho u) + \frac{\partial}{\partial y}(\rho v) + \frac{\partial}{\partial z}(\rho w) = 0 \tag{6.3}$$

Eqs. (6.1) and (6.2) can be transformed to a more convenient form. If we multiply (6.1) by ρ and (6.3) by u then add the resulting equations, we obtain the *flux* form of the x-momentum equation:

$$\frac{\partial}{\partial t}(\rho u) + \frac{\partial}{\partial x}(\rho u^2) + \frac{\partial}{\partial y}(\rho uv) + \frac{\partial}{\partial z}(\rho uw) - f\rho v = -\frac{\partial p}{\partial x} \tag{6.4}$$

An analogous operation gives the flux form of the y-momentum equation:

$$\frac{\partial}{\partial t}(\rho v) + \frac{\partial}{\partial x}(\rho uv) + \frac{\partial}{\partial y}(\rho v^2) + \frac{\partial}{\partial z}(\rho vw) + f\rho u = -\frac{\partial p}{\partial y} \quad (6.5)$$

We next substitute in place of u, v, and w in (6.4) and (6.5) letting

$$u = \langle u \rangle + u', \quad v = \langle v \rangle + v', \quad w = \langle w \rangle + w' \quad (6.6)$$

If we neglect the small fluctuations in density associated with the turbulence, the resulting equations can be averaged in time to get a simple partition of the flow between the mean flow and turbulent fields. Thus, for example, the term ρuw becomes

$$\langle \rho uw \rangle = \rho \langle (\langle u \rangle + u')(\langle w \rangle + w') \rangle = \rho \langle u \rangle \langle w \rangle + \rho \langle u'w' \rangle$$

where terms like $\langle \langle u \rangle w' \rangle$ and $\langle u' \langle w \rangle \rangle$ vanish because $\langle u \rangle$ and $\langle w \rangle$ are constant over the averaging interval so that, for example, $\langle \langle u \rangle w' \rangle = \langle u \rangle \langle w' \rangle = 0$ since $\langle w' \rangle = 0$. Carrying out this averaging process for all terms in (6.4) and (6.5) we obtain with the aid of the averaged continuity equation:

$$\frac{\partial \langle u \rangle}{\partial t} + \langle u \rangle \frac{\partial \langle u \rangle}{\partial x} + \langle v \rangle \frac{\partial \langle u \rangle}{\partial y} + \langle w \rangle \frac{\partial \langle u \rangle}{\partial z} - f \langle v \rangle$$

$$= -\frac{1}{\rho} \frac{\partial \langle p \rangle}{\partial x} - \frac{1}{\rho} \left[\frac{\partial}{\partial x}(\langle \rho u'u' \rangle) + \frac{\partial}{\partial y}(\langle \rho u'v' \rangle) + \frac{\partial}{\partial z}(\langle \rho u'w' \rangle) \right] \quad (6.7)$$

$$\frac{\partial \langle v \rangle}{\partial t} + \langle u \rangle \frac{\partial \langle v \rangle}{\partial x} + \langle v \rangle \frac{\partial \langle v \rangle}{\partial y} + \langle w \rangle \frac{\partial \langle v \rangle}{\partial z} + f \langle u \rangle$$

$$= -\frac{1}{\rho} \frac{\partial \langle p \rangle}{\partial y} - \frac{1}{\rho} \left[\frac{\partial}{\partial x}(\rho \langle u'v' \rangle) + \frac{\partial}{\partial y}(\rho \langle v'v' \rangle) + \frac{\partial}{\partial z}(\rho \langle v'w' \rangle) \right] \quad (6.8)$$

The terms in square brackets on the right-hand side in (6.7) and (6.8) which depend on the turbulent fluctuations are called the *eddy stress terms*.

In the mixing length theory these eddy stress terms are parameterized in terms of the mean field variables by assuming that the eddy stresses are proportional to the gradient of the mean wind. Since we are here primarily concerned with the planetary boundary layer where vertical gradients are much larger than the horizontal gradients, the discussion will be limited to the vertical eddy stress terms.

According to the mixing length hypothesis a parcel of fluid which is displaced vertically will carry the mean horizontal velocity of its original level for a characteristic distance l' analogous to the mean free path in molecular viscosity. This displacement will create a turbulent fluctuation whose magnitude will depend upon l' and the shear of the mean velocity. Thus, for example,

$$u' = -l' \frac{\partial \langle u \rangle}{\partial z}$$

where it must be understood that $l' > 0$ for *upward* parcel displacement and $l' < 0$ for *downward* parcel displacement. The vertical eddy stress $-\rho\langle u'w'\rangle$ can then be written as

$$-\rho\langle u'w'\rangle = \rho\langle w'l'\rangle \frac{\partial \langle u \rangle}{\partial z} \tag{6.9}$$

In order to estimate w' in terms of the mean fields we assume that the vertical stability of the atmosphere is nearly neutral so that buoyancy effects are small (see Section 9.4). The horizontal scale of the eddies should then be comparable to the vertical scale so that $|w'| \sim |\mathbf{V_h}'|$ and we can set

$$w' = l' \left| \frac{\partial \langle \mathbf{V_h} \rangle}{\partial z} \right|$$

where $\mathbf{V_h}'$ and $\langle \mathbf{V_h} \rangle$ designate the turbulent and mean parts of the horizontal velocity field, respectively. Here the absolute value of the mean velocity gradient is needed because if $l' > 0$ we must have $w' > 0$ (that is, upward parcel displacement by the eddy fluctuations). Thus the eddy stress can be written

$$-\rho\langle u'w'\rangle = \rho\langle l'^2 \rangle \left| \frac{\partial \langle \mathbf{V_h} \rangle}{\partial z} \right| \frac{\partial \langle u \rangle}{\partial z} = A_z \frac{\partial \langle u \rangle}{\partial z} \tag{6.10}$$

where $A_z \equiv \rho l'^2 |\partial \langle \mathbf{V_h} \rangle / \partial z|$ is called the *eddy exchange coefficient*. Similarly we may show that the vertical eddy stress due to motion in the y direction may be written

$$-\rho\langle v'w'\rangle = A_z \frac{\partial \langle v \rangle}{\partial z}$$

In the planetary boundary layer it is usually assumed that A_z depends only on the distance from the surface.

6.2 The Ekman Layer Equations

If the horizontal eddy stress terms in (6.7) and (6.8) are neglected and the vertical stresses are parameterized in terms of the mean flow by introducing the eddy exchange coefficient the momentum equations become

$$\frac{\partial u}{\partial t} + u\frac{\partial u}{\partial x} + v\frac{\partial u}{\partial y} + w\frac{\partial u}{\partial z} - fv = -\frac{1}{\rho}\frac{\partial p}{\partial x} + \frac{1}{\rho}\frac{\partial}{\partial z}\left(A_z\frac{\partial u}{\partial z}\right) \qquad (6.11)$$

$$\frac{\partial v}{\partial t} + u\frac{\partial v}{\partial x} + v\frac{\partial v}{\partial y} + w\frac{\partial v}{\partial z} + fu = -\frac{1}{\rho}\frac{\partial p}{\partial y} + \frac{1}{\rho}\frac{\partial}{\partial z}\left(A_z\frac{\partial v}{\partial z}\right) \qquad (6.12)$$

where we have omitted the angle brackets since all dependent variables are here mean flow variables. For midlatitude synoptic scale motions, we showed in Section 2.4 that the acceleration terms du/dt and dv/dt in (6.11) and (6.12) were small compared to the Coriolis force and pressure gradient force terms. Outside the boundary layer the first approximation was then simply geostrophic balance. In the boundary layer the acceleration terms are still small compared to the Coriolis and pressure gradient terms. Thus to a first approximation the planetary boundary layer equations express a three-way balance between the Coriolis force, the pressure gradient force, and the viscosity force:

$$-fv = -\frac{1}{\rho}\frac{\partial p}{\partial x} + \frac{1}{\rho}\frac{\partial}{\partial z}\left(A_z\frac{\partial u}{\partial z}\right) \qquad (6.13)$$

$$+fu = -\frac{1}{\rho}\frac{\partial p}{\partial y} + \frac{1}{\rho}\frac{\partial}{\partial z}\left(A_z\frac{\partial v}{\partial z}\right) \qquad (6.14)$$

To simplify further development we now assume that the eddy exchange coefficient is independent of height and introduce an eddy viscosity coefficient $K = A_z/\rho$ which is analogous to the molecular kinematic viscosity coefficient in laminar flow. Actually A_z cannot be constant all the way to the ground since the value of A_z depends on the mixing length which must decrease as one approaches the ground. However, the qualitative nature of the solutions is not changed by including a vertical dependence of A_z. The constant coefficient solution is also of interest because it can be modeled easily with laminar flow in the laboratory.

Thus, we obtain finally as our approximate equations for the planetary boundary layer:

$$K\frac{\partial^2 u}{\partial z^2} + f(v - v_g) = 0 \qquad (6.15)$$

$$K\frac{\partial^2 v}{\partial z^2} - f(u - u_g) = 0 \qquad (6.16)$$

where we have used the definitions

$$u_g \equiv -\frac{1}{\rho f}\frac{\partial p}{\partial y}, \qquad v_g \equiv \frac{1}{\rho f}\frac{\partial p}{\partial x}.$$

The Ekman layer equations (6.15) and (6.16) can be solved to determine the departure of the wind field from geostrophic balance in the boundary layer. To solve this set of simultaneous second-order differential equations we must specify boundary conditions on u and v. At the surface of the earth the velocity must go to zero, whereas in the free atmosphere sufficiently far from the surface the wind must approach its geostrophic value. Therefore, as boundary conditions on (6.15) and (6.16) we choose

$$u = 0, \quad v = 0 \quad \text{at} \quad z = 0$$
$$u \to u_g, \quad v \to v_g \quad \text{as} \quad z \to \infty \tag{6.17}$$

To solve this set it is convenient to multiply (6.16) by $i \equiv \sqrt{-1}$ and add the result to (6.15) to obtain a second-order equation in the *complex velocity* $u + iv$:

$$K \frac{\partial^2}{\partial z^2} (u + iv) - if(u + iv) = -if(u_g + iv_g) \tag{6.18}$$

For simplicity, we assume that the flow is oriented so that the geostrophic wind is entirely in the zonal direction ($v_g = 0$). Then the general solution of (6.18) may be written

$$u + iv = A \exp[(if/K)^{1/2}z] + B \exp[-(if/K)^{1/2}z]$$

It can be shown that $\sqrt{i} = (1 + i)/\sqrt{2}$. Using this relationship and applying the boundary conditions (6.17), we find that $A = 0$ and $B = -u_g$. Thus,

$$u + iv = -u_g e^{-\gamma(1+i)z} + u_g$$

where

$$\gamma = (f/2K)^{1/2}$$

Applying the Euler formula $e^{-i\theta} = \cos\theta - i\sin\theta$ and separating the real from the imaginary part we obtain

$$u = u_g(1 - e^{-\gamma z}\cos\gamma z)$$
$$v = u_g e^{-\gamma z}\sin\gamma z \tag{6.19}$$

This solution is the famous *Ekman spiral* named for the Swedish oceanographer V. W. Ekman who first derived an analogous solution for the surface wind drift current in the ocean. The structure of the solution (6.19) is best illustrated by a hodograph as shown in Fig. 6.1. In this figure the components of the wind velocity are plotted as a function of height. Thus the points on the curve in Fig. 6.1 correspond to u and v in (6.19) for values of γz increasing as one moves away from the origin along the spiral. It can be seen from Fig. 6.1 that when $z = \pi/\gamma$, the wind is parallel to the geostrophic wind although

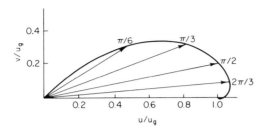

Fig. 6.1 Hodograph of the Ekman spiral solution. Points marked on the curve are values of γz which is a nondimensional measure of height.

slightly greater than geostrophic in magnitude. It is conventional to designate this level as the top of the planetary boundary layer. Thus the depth of the *Ekman layer* is

$$De \equiv \pi/\gamma \qquad (6.20)$$

Observations indicate that the wind approaches its geostrophic value at about one kilometer above the ground. Substituting $De = 1$ km and $f = 10^{-4}$ into (6.20), we can solve for the eddy viscosity K. The result is that $K \simeq 5 \times 10^4$ cm^2 sec^{-1}. Referring back to (6.10) we see that $K \simeq \langle l'^2 \rangle \, |\partial \langle \mathbf{V}_h \rangle / \partial z|$. Thus if the mean wind shear is of the order of 5 m sec^{-1} km^{-1}, the mixing length l' must be about 30 m in order that $K \simeq 5 \times 10^4$ cm^2 sec^{-1}. This mixing length is small compared to the depth of the boundary layer, as it should be if the mixing length concept is to be useful.

Qualitatively the most striking feature of the Ekman layer solution is the fact that the wind in the boundary layer has a component directed toward lower pressure. This is a direct result of the three-way balance between the pressure gradient force, the Coriolis force, and the viscous force. This balance is illustrated in Fig. 6.2 for a level well within the boundary layer. Since the Coriolis force is always normal to the velocity and the frictional force is mainly a retarding force, their sum can only exactly balance the pressure

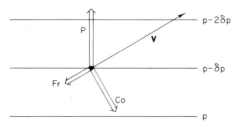

Fig. 6.2 Balance of forces in the Ekman layer. P designates the pressure gradient, Co designates the Coriolis force, Fr designates the frictional force.

gradient force if the wind is directed toward lower pressure (that is, to the left of the geostrophic wind in the northern hemisphere). Furthermore, it is easy to see that as the frictional force becomes increasingly important the cross isobar angle must increase.

The ideal Ekman layer discussed here is rarely, if ever, observed in the atmospheric boundary layer partly because, as mentioned above, the eddy mixing coefficient must vary rapidly with height near the ground. An additional reason, however, is that the Ekman layer wind profile is generally unstable for a neutrally buoyant atmosphere. The circulations which develop as a result of this instability have horizontal and vertical scales comparable to the depth of the boundary layer. Thus, it is not possible to parameterize them by a simple mixing length theory. However, these circulations do in general transport considerable momentum vertically. The net result is usually to decrease the angle between the boundary layer wind and the geostrophic wind. A typical observed wind hodograph is shown in Fig. 6.3. Although the detailed structure is rather different from the Ekman spiral, the vertically integrated horizontal mass transport in the boundary layer is still directed toward lower pressure. And as we shall see in the next section it is this fact which is of primary importance for synoptic scale systems.

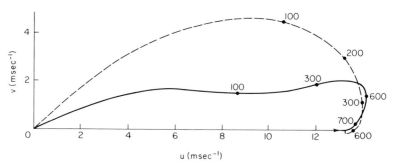

Fig. 6.3 Mean wind hodograph for Jacksonville, 4 April 1968 (solid line) compared with the Ekman spiral (dashed line). (Adapted from Brown, 1970.)

6.3 Secondary Circulations and Spin-Down

In the idealized Ekman spiral solution of (6.19) the v component of the wind multiplied by ρ gives the cross isobaric mass transport per unit area at any level in the boundary layer. Thus the net mass transport toward lower pressure in the Ekman layer for a column of unit width extending vertically through the entire layer is

$$M = \int_0^{De} \rho v \, dz = \int_0^{De} \rho u_g e^{-\pi z/De} \sin \frac{\pi z}{De} \, dz \qquad (6.21)$$

Neglecting local variations of density in the boundary layer the continuity equation may be written as

$$\frac{\partial}{\partial z}(\rho w) = -\frac{\partial}{\partial x}(\rho u) - \frac{\partial}{\partial y}(\rho v) \tag{6.22}$$

Integrating (6.22) through the depth of the boundary layer we find that

$$(w\rho)_{De} = -\int_0^{De}\left[\frac{\partial}{\partial x}(\rho u) + \frac{\partial}{\partial y}(\rho v)\right]dz$$

Here we have assumed that the ground is level so that $w = 0$ at $z = 0$. Substituting from (6.19) we can rewrite this expression for the vertical mass flux at the top of the Ekman layer as

$$(\rho w)_{De} = -\frac{\partial}{\partial y}\int_0^{De}\rho u_{\mathrm{g}}e^{-\pi z/De}\sin\frac{\pi z}{De}\,dz \tag{6.23}$$

Comparing (6.23) with (6.21) we see that the vertical flux at the top of the boundary layer is equal to the horizontal convergence of mass in the boundary layer which is simply $-\partial M/\partial y$ in the above example. Noting that $-\partial u_{\mathrm{g}}/\partial y \equiv \zeta_{\mathrm{g}}$ is just the geostrophic vorticity in this case, we have after integrating (6.23)

$$w(De) = \zeta_{\mathrm{g}}(K/2f)^{1/2} \tag{6.24}$$

where we have neglected the variation of density with height in the boundary layer. Hence, we obtain the important result that the vertical velocity at the top of the boundary layer is proportional to the geostrophic vorticity. In this way the effect of friction in the boundary layer is communicated directly to the interior fluid through a forced *secondary circulation* rather than indirectly by the slow process of viscous diffusion. For a typical synoptic scale system with $\zeta_{\mathrm{g}} \sim 10^{-5}\ \mathrm{sec}^{-1}, f \sim 10^{-4}\ \mathrm{sec}^{-1}$, and $De \sim 1$ km, the vertical velocity given by (6.24) is of the order of a few tenths of a centimeter per second.

An analogous secondary circulation is responsible for the decay of the circulation created when a cup of tea is stirred. Away from the boundary of the cup there is an approximate balance between the radial pressure gradient and the centrifugal force of the spinning fluid. However, near the bottom of the cup viscosity slows the motion and the centrifugal force is not sufficient to balance the radial pressure gradient. (Note that the radial pressure gradient is independent of depth since tea is an incompressible fluid.) Therefore, radial inflow takes place near the bottom of the cup. Because of this inflow the tea leaves always are observed to cluster near the center at the bottom of the cup if the tea has been stirred. By continuity the radial inflow in the bottom boundary layer requires upward motion and a slow compensating outward radial flow throughout the remaining portion of the cup. This slow outward flow approximately conserves angular momentum, and by replacing high angular momentum fluid by low angular momentum fluid serves to *spin-down* the vorticity in the cup far more rapidly than could mere molecular diffusion.

This spin-down effect is also important in the atmosphere. It is most easily illustrated in the case of a barotropic atmosphere. We showed previously in Section 5.5 that for synoptic scale motions the vorticity equation could be written approximately as

$$\frac{d}{dt}(\zeta + f) = -f\left(\frac{\partial u}{\partial x} + \frac{\partial v}{\partial y}\right) = f\frac{\partial w}{\partial z} \qquad (6.25)$$

where we have neglected ζ compared to f in the divergence term. Neglecting the latitudinal variation of f we next evaluate the integral of (6.25) from the top of the boundary layer, $z = De$, to the tropopause, $z = H$.

$$\int_{De}^{H} \frac{d\zeta}{dt}\,dz = f\int_{w(De)}^{w(H)} dw \qquad (6.26)$$

Assuming that $w = 0$ at $z = H$ and that the vorticity may be approximated by its geostrophic value (which is independent of height) we obtain from (6.26)

$$\frac{d\zeta_g}{dt} = \frac{-f}{(H - De)}\,w(De)$$

Substituting from (6.24) and noting that $H \gg De$ we obtain a differential equation for the time dependence of ζ_g :

$$\frac{d\zeta_g}{dt} = -\left(\frac{fK}{2H^2}\right)^{1/2}\zeta_g \qquad (6.27)$$

Equation (6.27) may be immediately integrated to give

$$\zeta_g = \zeta_g(0)\exp\{-(fK/2H^2)^{1/2}t\} \qquad (6.28)$$

where $\zeta_g(0)$ is the value of the geostrophic vorticity at time $t = 0$. From (6.28) we see that $\tau_e \equiv H(2/fK)^{1/2}$ is the time which it takes a barotropic vortex of height H to spin-down to e^{-1} of its original value (this "e-folding" time scale is what is meant by the *spin-down time*). Taking typical values of the parameters as follows: $H = 10$ km, $f = 10^{-4}$ sec^{-1}, and $K = 10^5$ cm^2 sec^{-1}, we find that $\tau_e \approx 4$ days. Thus, for midlatitude synoptic scale disturbances in a barotropic atmosphere the characteristic spin-down time is a few days. This decay time scale should be compared to the time scale for ordinary viscous diffusion. It can be shown that the time for eddy diffusion to penetrate a depth H is of the order $\tau_d \sim H^2/K$. For the above values of H and K, the diffusion time scale is thus $\tau_d \sim 100$ days. Hence, the spin-down process is a far more effective mechanism for destroying vorticity in a rotating atmosphere than is ordinary diffusion.

Physically the spin-down process in the atmospheric case is similar to that described for the teacup, except that in synoptic scale systems it is primarily the Coriolis force which balances the pressure gradient force away from the boundary, not the centrifugal force. Again the role of the secondary circulation driven by viscous forces in the boundary layer is to provide a slow radial flow in the interior which is superposed on the azimuthal circulation of the vortex. This secondary circulation is directed outward in a cyclone so that the horizontal area enclosed by any chain of fluid particles gradually increases. Since the circulation is conserved, the azimuthal velocity at any distance from the vortex center must decrease in time. Or, from another point of view, the Coriolis force for the outward-flowing fluid is directed clockwise, and this force thus exerts a torque opposite to the direction of the circulation of the vortex. In Fig. 6.4 a qualitative sketch of the streamlines of this secondary flow is shown.

It should now be obvious exactly what is meant by the term *secondary circulation*. It is simply a circulation superposed on the primary circulation

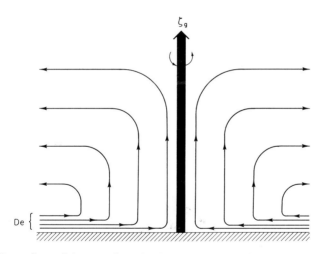

Fig. 6.4   Streamlines of the secondary circulation forced by frictional convergence in the planetary boundary layer for a cyclonic vortex in a barotropic atmosphere.

(in this case the azimuthal circulation of the vortex) by the physical constraints of the system. In the case of the boundary layer it is viscosity which is responsible for the presence of the secondary circulation. However, other processes such as temperature advection and diabatic heating may also lead to secondary circulations as we shall see later.

The above discussion has concerned only the neutrally stratified barotropic atmosphere. An analysis for the more realistic case of a stably stratified baroclinic atmosphere would be much more complicated. However, qualitatively the effects of stratification may be easily understood. The buoyancy force (see Section 9.4) will act to suppress vertical motion since air lifted vertically in a stable environment will be denser than the environmental air. As a result the interior secondary circulation will be restricted in vertical extent as shown in Fig. 6.5. Most of the return flow will take place just above the

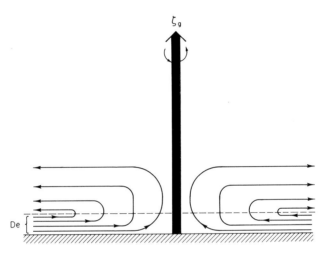

Fig. 6.5 Streamlines of the secondary circulation forced by frictional convergence in the planetary boundary layer for a cyclonic vortex in a stably stratified baroclinic atmosphere.

boundary layer. This secondary flow will rather quickly spin-down the vorticity at the top of the Ekman layer without appreciably affecting the higher levels. When the geostrophic vorticity at the top of the boundary layer is reduced to zero, the "pumping" action of the Ekman layer is eliminated. The result is a baroclinic vortex with a vertical shear of the azimuthal velocity which is just strong enough to bring ζ_g to zero at the top of the boundary layer. This vertical shear of the geostrophic wind requires a radial temperature gradient to satisfy the thermal wind relationship. This radial temperature gradient is in fact produced during the spin-down phase by the adiabatic cooling of the air forced out of the Ekman layer. Thus, the secondary circulation in the baroclinic atmosphere serves two purposes: (1) it changes the azimuthal velocity field of the vortex through the action of the Coriolis force, and (2) it changes the temperature distribution so that a thermal wind balance is always maintained between the vertical shear of the azimuthal velocity and the radial temperature gradient.

Problems

1. Verify by direct substitution that the Ekman spiral expression (6.19) is indeed a solution of the boundary layer equations (6.15) and (6.16).

2. Derive the Ekman spiral solution for the more general case where the geostrophic wind has both x and y components, u_g and v_g.

3. Letting the Coriolis parameter and density be constants, show that Equation (6.24) is correct for the more general Ekman spiral solution obtained in Problem 2.

4. For laminar flow in a rotating cylindrical vessel filled with water (molecular kinematic viscosity $v = 0.01$ cm^2 sec^{-1}), compute the depth of the Ekman layer and the spin-down time if the depth of the fluid is 30 cm and the rotation rate of the tank is ten revolutions per minute.

5. For the situation of Problem 4, how small would the radius of the tank have to be in order that the time scale for viscous diffusion from the sidewalls be comparable to the spin-down time?

6. In a homogeneous barotropic fluid show that the secondary radial circulation associated with a geostrophic vortex does not vary in strength with height.

7. Derive an expression for the wind driven surface Ekman layer in the ocean. Assume that the wind stress τ_w is constant and directed along the x-axis. Continuity of stress at the air–sea interface ($z = 0$) requires that the wind stress equals the water stress so that the boundary condition at the surface becomes

$$K\frac{\partial u}{\partial z} = \tau_w, \qquad K\frac{\partial v}{\partial z} = 0 \qquad \text{at } z = 0$$

where K is the eddy viscosity in the ocean (assumed constant). As a lower boundary condition assume that $u, v \to 0$ as $z \to -\infty$. If $K = 10$ cm^2 sec^{-1} what is the depth of the surface Ekman layer?

8. Show that the vertically integrated mass transport in the wind driven oceanic surface Ekman layer is directed 90° to the right of the surface wind stress in the Northern Hemisphere. Explain this result physically.

Suggested References

Batchelor, *An Introduction to Fluid Dynamics*, has a useful discussion of the molecular basis of viscosity as well as a succinct derivation of the complete stress tensor. This book also has an excellent treatment of viscous boundary layers including the Ekman layer.

Cole, *Perturbation Methods in Applied Mathematics* provides an excellent graduate level introduction to the mathematical foundations of boundary layer theory.

Greenspan, *The Theory of Rotating Fluids* gives a unified account of the role of viscosity in rotating fluids at an advanced level. This is the only book available which focuses solely on phenomena which occur only in rotating fluids.

Lumley and Panofsky, *The Structure of Atmospheric Turbulence* covers both observational and theoretical aspects of the subject including the effects of thermal stratification in the boundary layer.

7 | Diagnostic Analysis of Synoptic Scale Motions in Middle Latitudes

A primary goal of dynamic meteorology is to interpret the observed structure of large-scale atmospheric motion systems in terms of the physical laws governing the motions. In Chapter 3 we discussed the constraints imposed on horizontal motions by Newton's second law, while in Chapter 4 the constraints imposed on vertical motions by the requirement of mass continuity were discussed. These two laws are not in themselves adequate to define completely the relationship between the mass and velocity fields in atmospheric circulation systems. In addition, we must utilize the law of energy conservation. In this chapter we will show from scaling considerations that the laws of motion, mass continuity, and energy conservation constrain synoptic scale disturbances so that to a good approximation *the three-dimensional velocity field is uniquely determined by the geopotential field.*

We first discuss the observed structure of midlatitude synoptic systems and the mean circulations in which they are embedded. We then develop two diagnostic equations, the geopotential tendency equation and the "omega" equation. Together, these equations enable us to diagnose the three-dimensional structure of a synoptic system. Finally, we shall use these diagnostic relationships as aids in obtaining an idealized model for a typical developing synoptic disturbance.

7.1   The Observed Structure of Midlatitude Synoptic Systems

In the real atmosphere circulation systems plotted on a synoptic map rarely resemble the simple circular vortices discussed in Chapter 3. Rather, they are generally highly asymmetric in form with the highest winds and largest temperature gradients concentrated along narrow bands called *fronts*. Also, such systems generally are highly baroclinic with both the amplitudes and phases of the geopotential and velocity perturbations changing substantially with height. Part of this complexity is due to the fact that these synoptic systems are not superposed on a uniform mean flow, but are embedded in a planetary scale flow which is itself highly baroclinic. Furthermore, this planetary scale flow is influenced by *orography* (that is, by large-scale terrain variations) and continent–ocean heating contrasts so that it is highly longitude dependent. Therefore, it is not accurate to view synoptic systems as disturbances superposed on a zonal flow which varies only with latitude and height. However, as we shall see in Chapter 10, such a point of view is very useful as a first approximation in theoretical analysis of synoptic wave disturbances.

Indeed, zonally averaged cross sections do provide some useful information on the gross structure of the planetary scale circulation. In Fig. 7.1 we show mean meridional cross sections for the longitudinally averaged flow in the Northern Hemisphere during the months of January and July. As would be expected the pole-to-equator temperature gradient in the troposphere is much larger in the winter than in the summer. Since the zonal wind and temperature fields satisfy the thermal wind relationship (3.30) to a high degree of accuracy, the maximum zonal wind speed is much larger in the winter than in the summer. Furthermore, in both seasons the core of maximum zonal wind speed (called the jet stream axis) is located just below the tropopause at the height where the meridional temperature gradient vanishes and at the latitude where the average meridional temperature gradient in the troposphere divided by the Coriolis parameter is a maximum.

That the zonally averaged meridional cross sections of Fig. 7.1 are not representative of the mean wind structure at all longitudes can be seen in Fig. 7.2 which shows meridional cross sections of the zonal wind averaged over the months December–February at two widely separated longitudes in the Northern Hemisphere. The explanation for the large differences in the jet stream structure at these two longitudes is readily seen from examination of Fig. 7.3 which shows the mean 500-mb geopotential contours for January in the Northern Hemisphere. Even after averaging the height field for a month, very striking departures from zonal symmetry remain. These are clearly linked to the distributions of continents and oceans. The most prominent asymmetries are the troughs to the east of the American and Asian

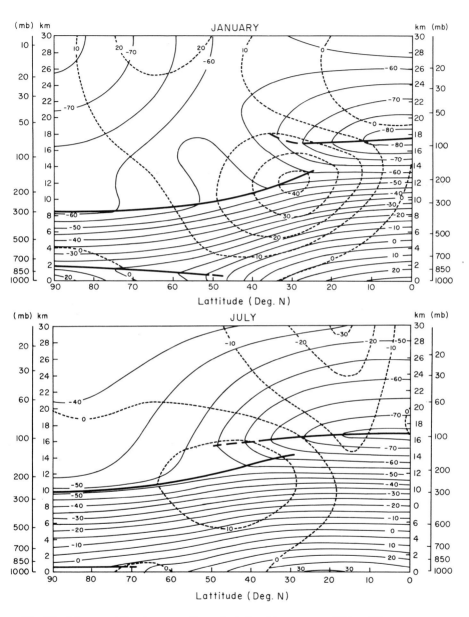

Fig. 7.1 Mean meridional cross sections of wind and temperature. Thin solid temperature lines in degrees centigrade. Dashed wind lines in meters per second. Heavy solid lines represent tropopause and inversion discontinuities. (After *Arctic Forecast Guide*, Navy Weather Research Facility, April 1962.)

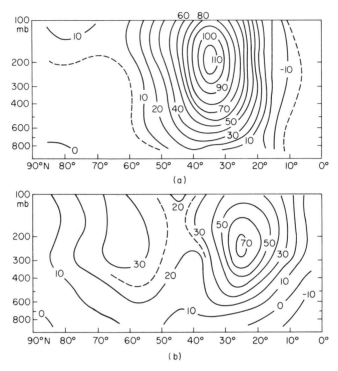

Fig. 7.2 Winter mean zonal winds in the Northern Hemisphere at (a) 140°E and (b) 0° longitude. Speeds shown are in knots (2 knots approximately equals 1 m sec^{-1}). (After Palmén and Newton, 1969.)

continents. Referring back to Fig. 7.2 we see that the intense jet at 35°N in the 140°E cross section is a result of the semipermanent trough at that longitude. Thus, it is apparent that the mean flow in which the synoptic systems are embedded should really be regarded as longitude dependent.

In addition to its longitudinal dependence the planetary scale flow also varies somewhat from day to day due to its interactions with transient synoptic scale disturbances. As a result, monthly mean charts tend to smooth out the actual structure of the jet stream since the position and intensity of the jet vary. Thus, at any time the planetary scale flow in the region of the jet stream has much greater baroclinicity than indicated on time-averaged charts. This point is illustrated schematically in Fig. 7.4 which shows an idealized vertical cross section through the jet stream. At any instant the axis of the jet stream tends to coincide with a narrow zone of strong temperature gradients called the *polar front*. This is the zone which in general separates the cold air of polar origin from warm tropical air. The occurrence of an intense jet core above the frontal zone is, of course, not mere coincidence, but rather a consequence of the thermal wind balance.

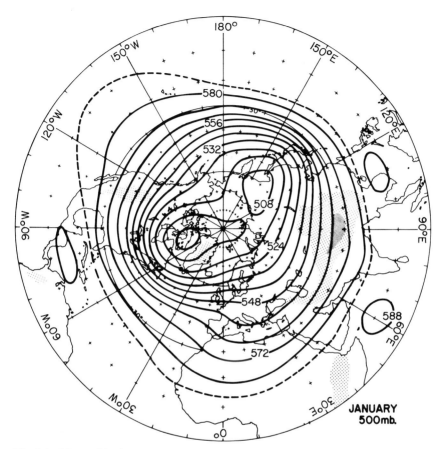

Fig. 7.3 Mean 500-mb contours in January, Northern Hemisphere. Heights shown in tens of meters. (After Palmén and Newton, 1969.)

It is a common observation in fluid dynamics that jet flows in which strong velocity shears occur may be unstable with respect to small perturbations. By this is meant that any small disturbance introduced into the jet flow will tend to amplify, drawing energy from the jet as it grows. Most synoptic scale systems in midlatitudes appear to develop as the result of an instability of the jet stream flow. This instability, called *baroclinic instability*, depends primarily on vertical shear in the jet stream, and hence tends to occur primarily in the region of the frontal zone. Baroclinic instability is not, however, identical to frontal instability since most baroclinic instability models describe only geostrophically scaled motions while disturbances in the vicinity of strong frontal zones must be highly nongeostrophic. The connection between conventional baroclinic instability and frontal instability is not yet completely understood.

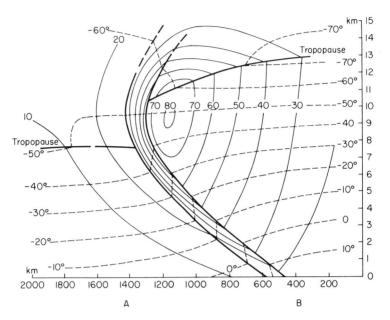

Fig. 7.4   Schematic isotherms (dashed lines, degrees centigrade) and isotachs (thin solid lines, meters per seond) in the vicinity of the polar front. Heavy lines indicate boundaries of the frontal zone and tropopauses. (After Palmén and Newton, 1969.)

The stages in the development of a typical baroclinic cyclone which develops as a result of baroclinic instability along the polar front are shown schematically in Fig. 7.5. In the stage of rapid development strong cold advection is seen to occur west of the trough at the surface, with weaker warm advection to the east. This pattern of thermal advection is a direct consequence of the fact that the trough at 500 mb lags (lies to the west of) the surface trough so that the mean geostrophic wind in the 500–1000-mb layer is directed across the 1000–500-mb thickness lines toward larger thickness west of the surface trough, and toward small thickness east of the trough. This dependence of the phase of the disturbance on height is better illustrated by Fig. 7.6 which shows a schematic downstream (or west–east) cross section through a developing baroclinic system. Throughout the troposphere the trough and ridge axes tilt westward (or upstream) with height while, consistent with the thermal wind relationship, the axes of warmest and coldest air have the opposite tilt. As we shall see later, these phase relationships are necessary in order that the mean flow give up potential energy to the developing wave. In the mature stage shown in the lower part of Fig. 7.5 the troughs at 500 and 1000 mb are nearly in phase. As a consequence, the thermal advection and energy conversion are quite weak.

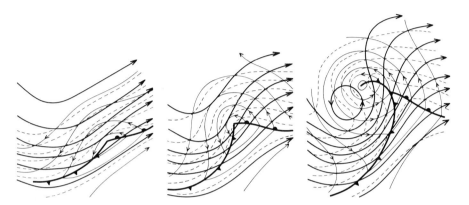

Fig. 7.5 Schematic 500-mb contours (heavy solid lines), 1000-mb contours (thin lines), and 1000–500-mb thickness (dashed) for a developing baroclinic wave. (After Palmén and Newton, 1969.)

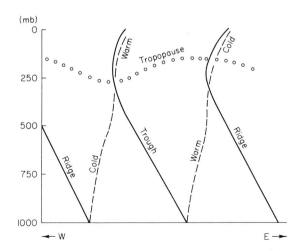

Fig. 7.6 West–east cross section through developing baroclinic wave. Solid lines are trough and ridge axes, dashed lines are axes of temperature extrema, chain of open circles denotes the tropopause.

7.2 The Basic Equations in Isobaric Coordinates

In the remainder of this chapter we wish to show from scaling considerations that the observed structure of midlatitude synoptic systems may be understood to be a consequence of the constraints imposed on the motions by Newton's second law, mass continuity, and energy conservation. For this purpose, it is convenient to develop the governing equations which are the mathematical expressions for these laws using the isobaric coordinate system. As pointed out in Chapter 3, meteorological measurements are generally referred to constant-pressure surfaces. In addition, the continuity equation takes a simple form in isobaric coordinates. Thus, use of the isobaric coordinate system will simplify the development of diagnostic equations.

We first consider the horizontal equations of motion. In Section 2.4.2 we showed that above the friction layer the approximate horizontal momentum equations for synoptic scale motions may be written as

$$\frac{du}{dt} - fv = -\frac{1}{\rho}\frac{\partial p}{\partial x} \tag{7.1}$$

$$\frac{dv}{dt} + fu = -\frac{1}{\rho}\frac{\partial p}{\partial y} \tag{7.2}$$

Following the method of Section 3.1, we transform the pressure gradient terms into geopotential gradients on an isobaric surface to obtain

$$\frac{du}{dt} - fv = -\frac{\partial \Phi}{\partial x} \tag{7.3}$$

$$\frac{dv}{dt} + fu = -\frac{\partial \Phi}{\partial y} \tag{7.4}$$

where the velocities are now measured with respect to isobaric surfaces and the derivative following the motion is defined according to

$$\frac{d}{dt} \equiv \left(\frac{\partial}{\partial t}\right)_p + u\left(\frac{\partial}{\partial x}\right)_p + v\left(\frac{\partial}{\partial y}\right)_p + \omega\frac{\partial}{\partial p} \tag{7.5}$$

where $\omega \equiv dp/dt$ is the individual pressure change.

The hydrostatic approximation and the continuity equation for isobaric coordinates have previously been discussed in Sections 3.4 and 4.2, respectively. The hydrostatic approximation may be written as

$$\frac{\partial \Phi}{\partial p} = -\alpha \tag{7.6}$$

and the continuity equation as

$$\frac{\partial u}{\partial x} + \frac{\partial v}{\partial y} + \frac{\partial \omega}{\partial p} = 0 \tag{7.7}$$

The thermodynamic energy equation can be written in its entropy form following the discussion in Section 1.5 as

$$c_p \frac{d \ln \theta}{dt} = \frac{dQ}{dt} \tag{7.8}$$

where the operator d/dt is defined by (7.5), and the potential temperature θ was defined in (1.18). The sixth, and final, governing equation is obtained by using the ideal gas law to eliminate T in (1.18) and obtain a relationship between θ and α in terms of the independent variable p:

$$\theta = \frac{p\alpha}{R} \left(\frac{1000}{p} \right)^{R/c_p} \tag{7.9}$$

Equations (7.3), (7.4), and (7.6)–(7.9) are thus a closed set which completely determines the relationships among the dependent variables u, v, ω, Φ, α, and θ provided that dQ/dt is specified.

7.3 Diagnostic Equations

The set of equations derived in the previous section, although already considerably simplified, is still very awkward to use for acquiring an understanding of the structure of synoptic scale systems. In this section we will show how scale analysis may be used to derive a much simplified set of equations suitable for diagnostic analysis of synoptic systems. In fact, we will show that for midlatitude synoptic scale systems the fields of geopotential tendency and vertical motion (that is, $\partial \Phi/\partial t$ and ω) are uniquely determined for any given distribution of geopotential Φ.

In order to obtain this set of diagnostic equations, we first eliminate θ by rewriting the thermodynamic energy equation in terms of Φ. This transformation of variables is possible because for an atmosphere in hydrostatic equilibrium $\partial \Phi/\partial p$ is proportional to the temperature. Taking the logarithm of (7.9) we get

$$\ln \theta = \ln \alpha - \left(\frac{R}{c_p} - 1 \right) \ln p + \text{const}$$

so that differentiation on an isobaric surface gives

$$\left(\frac{\partial \ln \theta}{\partial x} \right)_p = \left(\frac{\partial \ln \alpha}{\partial x} \right)_p, \qquad \text{etc.}$$

Using (7.5) to expand the total derivative $d \ln \theta / dt$ in (7.8) and substituting $\ln \alpha$ in place of $\ln \theta$ in the partial derivatives evaluated at constant p we get

$$\frac{\partial \ln \alpha}{\partial t} + u \frac{\partial \ln \alpha}{\partial x} + v \frac{\partial \ln \alpha}{\partial y} + \omega \frac{\partial \ln \theta}{\partial p} = \frac{1}{c_p} \frac{dQ}{dt}$$

Substituting from the hydrostatic approximation (7.6) to eliminate α in the terms on the left gives

$$\frac{\partial}{\partial t}\left(-\frac{\partial \Phi}{\partial p}\right) + u \frac{\partial}{\partial x}\left(-\frac{\partial \Phi}{\partial p}\right) + v \frac{\partial}{\partial y}\left(-\frac{\partial \Phi}{\partial p}\right) - \sigma \omega = \frac{\alpha}{c_p} \frac{dQ}{dt} \qquad (7.10)$$

where σ, the *static stability parameter*, is defined by

$$\sigma \equiv -\frac{\alpha}{\theta} \frac{\partial \theta}{\partial p}$$

For a statically stable atmosphere $\partial \theta / \partial p < 0$ so that $\sigma > 0$.[1]

Equation (7.10) may be further simplified if we note that for synoptic scale systems the horizontal velocity is approximately equal to the geostrophic velocity[2]

$$\mathbf{V} = \mathbf{i}u + \mathbf{j}v \simeq \mathbf{V}_g \equiv \frac{\mathbf{k} \times \nabla \Phi}{f}$$

Thus, to a first approximation, the horizontal velocity components in (7.10) may be replaced by their geostrophic values so that in vectorial notation

$$u \frac{\partial}{\partial x}\left(-\frac{\partial \Phi}{\partial p}\right) + v \frac{\partial}{\partial y}\left(-\frac{\partial \Phi}{\partial p}\right) \simeq \mathbf{V}_g \cdot \nabla\left(-\frac{\partial \Phi}{\partial p}\right)$$

Further, we assume that the diabatic heating dQ/dt is small compared to the terms on the left in (7.10). The approximate thermodynamic energy equation is thus

$$\frac{\partial}{\partial t}\left(-\frac{\partial \Phi}{\partial p}\right) = -\mathbf{V}_g \cdot \nabla\left(-\frac{\partial \Phi}{\partial p}\right) + \sigma \omega \qquad (7.11)$$

It will be shown in Problem 1 that σ can be expressed in terms of Φ so that (7.11) contains only the two dependent variables Φ and ω.

In giving a physical interpretation to the terms in (7.11), we will refer to $-\partial \Phi / \partial p$ as the "temperature" since it is proportional to temperature in a

---

[1] If θ increases with height (that is, decreases with pressure) an air parcel in adiabatic motion will be positively (negatively) buoyant when displaced vertically downward (upward) so that it will tend to return to its equilibrium level and the atmosphere is said to be statically stable.

[2] In this and all subsequent chapters $\mathbf{V}$ and ∇ will denote the *horizontal* velocity and *horizontal* del operator, respectively.

hydrostatic atmosphere. Alternatively, we could call $-\partial\Phi/\partial p$ the "thickness" since it is equal to the thickness $\delta\Phi$ divided by the pressure interval δp in the limit $\delta p \to 0$. Thus, the term of the left-hand side in (7.11) is proportional to the local rate of change of temperature on an isobaric surface. Similarly, the first term on the right-hand side is proportional to the advection of temperature by the geostrophic wind on an isobaric surface. The second term on the right in (7.11) is usually called the adiabatic cooling (heating) term. This term expresses the adiabatic temperature changes which result from rising and expansion (sinking and compression) of air parcels in a stable environment. The diabatic heating rate is usually small compared to the horizontal temperature advection and adiabatic cooling terms for midlatitude synoptic scale motions. For simplicity, we shall generally neglect diabatic heating in subsequent discussions although for time scales of more than a couple of days it should be included.

To further simplify our system of equations, it is convenient to replace the horizontal equations of motion by the vorticity equation. In Section 5.5 we previously discussed the scale analysis of the vorticity equation in the x, y, z coordinate system. The discussion for isobaric coordinates is similar, although somewhat simplified because the solenoidal term is automatically excluded when horizontal derivatives are evaluated on isobaric surfaces.

To obtain an equation for the vertical component of vorticity in the isobaric coordinate system we differentiate (7.3) with respect to y, (7.4) with respect to x, and subtract the former from the latter to obtain

$$\frac{\partial\zeta}{\partial t} = -\mathbf{V}\cdot\mathbf{V}(\zeta+f) - \omega\frac{\partial\zeta}{\partial p} - (\zeta+f)\mathbf{V}\cdot\mathbf{V} + \left\{\frac{\partial u}{\partial p}\frac{\partial\omega}{\partial y} - \frac{\partial v}{\partial p}\frac{\partial\omega}{\partial x}\right\} \quad (7.12)$$

where $\zeta = \mathbf{k}\cdot(\mathbf{V}\times\mathbf{V})$ and all horizontal derivatives are evaluated at constant pressure. The terms in (7.12) in order reading from left to right are as follows:

1. the local rate of change of relative vorticity,
2. the horizontal advection of absolute vorticity,
3. the vertical advection of relative vorticity,
4. the divergence term,
5. the twisting or tilting term.

As shown by the scale analysis of Chapter 5, we may simplify the vorticity equation for midlatitude synoptic scale motions by:

1. neglecting the vertical advection and twisting terms;
2. neglecting ζ compared to f in the divergence term;
3. approximating the horizontal velocity by the geostrophic wind in the *advection term*;
4. replacing the relative vorticity by its geostrophic value.

As a further simplification, we may expand the Coriolis parameter in a Taylor series about the latitude ϕ_0 as

$$f = f_0 + \beta y + \text{(higher-order terms)}$$

where $\beta \equiv (df/dy)_{\phi_0}$, and $y = 0$ at ϕ_0. If we let L designate the latitudinal scale of the motions, then the ratio of the first two terms in the expansion of f has order of magnitude

$$\frac{\beta L}{f_0} \sim \frac{\cos \phi_0}{\sin \phi_0} \frac{L}{a}$$

Thus, when the latitudinal scale of the motions is small compared to the radius of the earth ($L/a \ll 1$) we can let the Coriolis parameter have a constant value f_0 except where it appears differentiated in the advection term in which case $df/dy \equiv \beta$ is assumed to be constant This approximation is usually referred to as the beta-plane approximation.

Applying all the above approximations, we obtain the *quasi-geostrophic vorticity equation*

$$\frac{\partial \zeta_g}{\partial t} = -\mathbf{V}_g \cdot \nabla(\zeta_g + f) - f_0 \nabla \cdot \mathbf{V} \qquad (7.13)$$

where $\zeta_g = \nabla^2 \Phi / f_0$ and $\mathbf{V}_g = \mathbf{k} \times \nabla \Phi / f_0$ are both evaluated using the constant Coriolis paramater f_0.

It is very important to note that the horizontal wind is *not* replaced by its geostrophic value in the divergence term. In fact, when the geostrophic wind is computed using a constant Coriolis parameter, it is just the small departures of the horizontal wind from geostrophy which account for the divergence. As we shall see later, this divergence and its corresponding vertical motion field are dynamically necessary to keep the temperature changes hydrostatic and vorticity changes geostrophic in synoptic scale systems.

The horizontal divergence in (7.13) can easily be eliminated using the continuity equation

$$\nabla \cdot \mathbf{V} = -\frac{\partial \omega}{\partial p}$$

to obtain an alternative form of the quasi-geostrophic vorticity equation

$$\frac{\partial \zeta_g}{\partial t} = -\mathbf{V}_g \cdot \nabla(\zeta_g + f) + f_0 \frac{\partial \omega}{\partial p} \qquad (7.14)$$

Since ζ_g and $\mathbf{V}_g$ are both defined in terms of Φ (7.14) can be used to diagnose the ω field provided that the fields of both Φ and $\partial \Phi / \partial t$ are known. Since both the local change of the geostrophic vorticity and the advection of vorticity by the geostrophic wind can be estimated with reasonable accuracy estimates of ω based on (7.14) usually are better than the estimates based on the continuity equation discussed in Chapter 4.

Because ζ_g and $\mathbf{V}_g$ are both functions of Φ the quasi-geostrophic vorticity equation (7.14) and the hydrostatic thermodynamic energy equation (7.11) each contain only the two dependent variables Φ and ω. Therefore, (7.11) and (7.14) form a closed set of prediction equations in Φ and ω. It is thus possible to eliminate ω between these two equations and obtain an equation relating Φ to $\partial\Phi/\partial t$. This equation is called the *geopotential tendency equation*. It is actually just a form of the potential vorticity equation (see Section 8.4). Alternatively, we can eliminate the time derivative terms between (7.11) and (7.14) and obtain an equation which relates the ω field at any instant to the Φ field at that instant. This equation is called the vertical motion or *omega equation*.

Thus, by suitable manipulation of (7.11) and (7.14) we can obtain expressions which allow us to compute the geopotential tendency $\partial\Phi/\partial t$ and the vertical motion ω from observations of the instantaneous field of Φ alone. Hence, to a first approximation, the evolution of midlatitude synoptic scale flow can be predicted without direct measurement of the velocity field. This set of equations, which really constitutes the core of modern dynamic meteorology, is called the *quasi-geostrophic system*.

7.3.1 THE TENDENCY EQUATION

If we define the geopotential tendency as $\chi \equiv \partial\Phi/\partial t$, then (7.11) and (7.14) may be written as

$$\frac{\partial\chi}{\partial p} = -\mathbf{V}_g \cdot \nabla\left(\frac{\partial\Phi}{\partial p}\right) - \sigma\omega \tag{7.15}$$

$$\nabla^2\chi = -f_0\mathbf{V}_g \cdot \nabla\left(\frac{1}{f_0}\nabla^2\Phi + f\right) + f_0{}^2\frac{\partial\omega}{\partial p} \tag{7.16}$$

where we have used the relationship $\zeta_g = (1/f_0)\nabla^2\Phi$ so that after changing the order of partial differentiation,

$$\frac{\partial\zeta_g}{\partial t} = \frac{1}{f_0}\nabla^2\chi$$

If we multiply (7.15) by $f_0{}^2/\sigma$, then differentiate with respect to pressure and add the result to (7.16), we obtain

$$\underbrace{\left(\nabla^2 + \frac{f_0{}^2}{\sigma}\frac{\partial^2}{\partial p^2}\right)\chi}_{A} = \underbrace{-f_0\mathbf{V}_g \cdot \nabla\left(\frac{1}{f_0}\nabla^2\Phi + f\right)}_{B} + \underbrace{\frac{f_0{}^2}{\sigma}\frac{\partial}{\partial p}\left(-\mathbf{V}_g \cdot \nabla\frac{\partial\Phi}{\partial p}\right)}_{C} \tag{7.17}$$

where we have assumed σ to be a constant.[3] Equation (7.17) is the geopotential tendency equation.

We shall now discuss in turn the physical interpretation of each term in (7.17). Term A involves only second derivatives in space of the χ field. For wavelike disturbances, this term can be shown to be proportional to $-\chi$. To demonstrate this fact, we assume that the fields of Φ and χ vary sinusoidally in x and y.[4] Thus we may write

$$\chi = X(p) \sin kx \sin ly \qquad (7.18)$$

where the *wave numbers* k and l are defined as

$$k = \frac{2\pi}{L_x}, \qquad l = \frac{2\pi}{L_y}$$

with L_x and L_y the wavelengths in the x and y directions, respectively. The horizontal Laplacian of χ is then simply

$$\nabla^2 \chi = -(k^2 + l^2)\chi \propto -\chi$$

Similarly, since it is observed that midlatitude synoptic systems generally have a depth scale comparable to the depth of the troposphere we can crudely approximate the vertical variation of χ by letting

$$\frac{\partial^2 X}{\partial p^2} \simeq -\left(\frac{\pi}{p_0}\right)^2 \chi$$

where $p_0 = 1000$ mb. Term A can now be written approximately

$$\left(\nabla^2 + \frac{f_0^2}{\sigma}\frac{\partial^2}{\partial p^2}\right)\chi \propto -\left[k^2 + l^2 + \frac{1}{\sigma}\left(\frac{f_0\pi}{p_0}\right)^2\right]\chi$$

So that the left side in (7.17) is just proportional to the *negative* of the geopotential tendency.

We next consider the first term on the right-hand side in (7.17). This term is proportional to the advection of absolute vorticity by the geostrophic wind. For physical interpretation, it is convenient to divide term B into two parts by writing

$$\mathbf{V}_g \cdot \nabla\left(\frac{1}{f_0}\nabla^2\Phi + f\right) = \mathbf{V}_g \cdot \nabla\left(\frac{1}{f_0}\nabla^2\Phi\right) + v_g\frac{df}{dy}$$

---

[3] Actually, σ varies substantially with pressure even in the troposphere. However, the qualitative discussion in this section would not be changed if we were to include this additional complication.

[4] This assumption is not as restrictive as it may appear because any bounded continuous function can be expanded in a double Fourier series in x and y. Thus our discussion would apply to any single component of the Fourier expansion for any distribution of Φ.

These two parts represent the geostrophic advections of relative vorticity and planetary vorticity, respectively. For disturbances in the westerlies, these two effects tend to oppose each other as illustrated by the 500-mb wave disturbance shown schematically in Fig. 7.7.

In region I upstream from the 500-mb trough, the geostrophic wind is directed from the negative vorticity maximum at the ridge toward the positive vorticity maximum at the trough so that

$$\mathbf{V}_g \cdot \mathbf{V} \left(\frac{1}{f_0} \, \nabla^2 \Phi \right) > 0$$

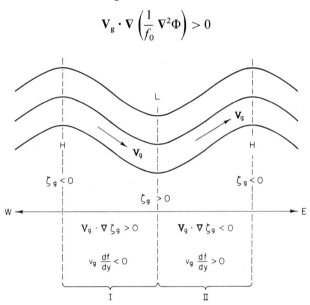

Fig. 7.7 Schematic 500-mb geopotential field showing regions of positive and negative advections of relative and planetary vorticity.

But at the same time since $v_g < 0$ in region I, the geostrophic wind has its y component directed down the gradient of planetary vorticity so that $v_g \, df/dy < 0$. Hence in region I the advection of relative vorticity tends to decrease the vorticity, whereas advection of planetary vorticity tends to increase the vorticity. Similar arguments (but with reversed signs) apply to region II. Therefore, advection of relative vorticity tends to move the vorticity pattern and hence the troughs and ridges eastward (downstream). But advection of planetary vorticity tends to move the troughs and ridges westward against the advecting wind field. This type of motion is called *retrograde* motion or *retrogression*.

The actual displacement of the pattern will obviously depend upon which type of vorticity advection dominates. In order to estimate the relative magnitudes of the relative vorticity and planetary vorticity advections, we assume that Φ has a sinusoidal dependence like that given for χ in (7.18). We can then write

$$\zeta_g = \frac{1}{f_0} \nabla^2 \Phi \simeq - \frac{(k^2 + l^2)}{f_0} \Phi$$

Thus, for a disturbance of given amplitude the relative vorticity increases for increasing wave number or *decreasing* wavelength. As a consequence, the advection of relative vorticity tends to dominate for short waves ($L_x \gtrsim 3000$ km) while for long waves ($L_x \gtrsim 10,000$ km) the planetary vorticity advection tends to dominate. Therefore, as a general rule short-wavelength synoptic scale systems should move rapidly eastward with the advecting zonal flow while long planetary waves should tend to retrograde.[5] Waves of intermediate wavelength may be quasi-stationary, or move eastward much slower than the mean geostrophic wind speed.

Returning to (7.17) we see that for short waves term B is negative in region I. Thus, as a result of vorticity advection χ will tend to be positive (recall that term A is proportional to $-\chi$). Hence, the geopotential height tendency will be positive and a ridge will tend to develop in this region. This ridging is, of course, necessary for the development of a negative geostrophic vorticity. Similar arguments, but with the signs reversed, apply to region II downstream from the trough where falling geopotential heights are associated with a positive relative vorticity advection. It is also important to note that the vorticity advection term is zero along both the trough and ridge axes since both $\nabla \zeta_g$ and v_g are zero at the axes. Thus, vorticity advection cannot change the strength of this type of disturbance, but only acts to propagate the disturbance horizontally.

The mechanism for amplification or decay of midlatitude synoptic systems is contained in term C of (7.17). This term, called the differential thickness advection, tends to be a maximum at the trough and ridge lines in a developing baroclinic wave. Now since

$$\mathbf{V}_g \cdot \mathbf{V} \left(\frac{\partial \Phi}{\partial p} \right)$$

is the thickess advection, which is proportional to the hydrostatic temperature advection, it is clear that

$$-\frac{\partial}{\partial p} \left[\mathbf{V}_g \cdot \mathbf{V} \left(\frac{\partial \Phi}{\partial p} \right) \right]$$

---

[5] The observed long waves in the atmosphere actually appear to be fixed in position rather than retrogade. This is believed to be a result of the forcing due to topographic influences and land continent heating contrasts as was previously mentioned in Section 7.1. However, detailed studies indicate that there do exist long-wave components of the motion which retrogade rapidly. In general these contain much less energy than the stationary component.

is proportional to the rate of change of temperature advection with height or the *differential temperature advection*.

To examine the influence of differential temperature advection on the geopotential tendency we consider the idealized developing wave shown in Fig. 7.5. Below the 500-mb ridge there is strong warm advection associated with the warm front, while below the 500-mb trough there is strong cold advection associated with the cold front. Now in the region of warm advection

$$V_g \cdot \nabla T < 0$$

$$\mathbf{V}_g \cdot \mathbf{V} \left(\frac{\partial \Phi}{\partial p} \right) > 0$$

since $\mathbf{V}_g$ has a component down the temperature gradient. But the warm advection decreases with height (increases with pressure) so that

$$\frac{\partial}{\partial p} \left[\mathbf{V}_g \cdot \mathbf{V} \left(\frac{\partial \Phi}{\partial p} \right) \right] > 0$$

Conversely, beneath the 500-mb trough where there is cold advection decreasing with height the opposite signs obtain. Therefore, along the 500-mb trough and ridge axes where the vorticity advection is zero the tendency equation states that for a developing wave

$$\chi \sim \frac{\partial}{\partial p} \left[\mathbf{V}_g \cdot \mathbf{V} \left(\frac{\partial \Phi}{\partial p} \right) \right] \begin{array}{l} > 0 \quad \text{at the ridge} \\ < 0 \quad \text{at the trough} \end{array}$$

Therefore, as indicated in Fig. 7.8, the effect of cold advection below the 500-mb trough is to *deepen* the trough, and the effect of warm advection below the 500-mb ridge is to *build* the ridge. Hence, it is the differential temperature or thickness advection which intensifies the upper-level troughs and ridges in a developing system.

Qualitatively the effects of differential temperature advection may be easily understood since the advection of cold air into the air column below the 500-mb trough will reduce the thickness of that column, and hence will lower the height of the 500-mb surface unless there is a compensating rise in the surface pressure. Obviously warm advection into the air column below the ridge will have the opposite effect.

In summary, we have shown that in the absence of diabatic heating the horizontal temperature advection must be non zero in order that a midlatitude synoptic system intensify through baroclinic processes. As we shall see later in Chapter 10 the temperature advection pattern described above implies conversion of potential energy to kinetic energy.

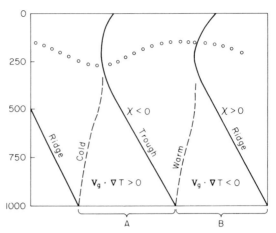

Fig. 7.8   East–west section through a developing synoptic disturbance showing the relationship of temperature advection to the upper level height tendencies. A and B designate, respectively, regions of cold advection and warm advection in the lower troposphere.

7.3.2   THE OMEGA EQUATION

A diagnostic equation for the vertical motion field may be obtained by eliminating χ between (7.15) and (7.16). To do this, we take the horizontal Laplacian of (7.15) to obtain

$$\nabla^2 \frac{\partial \chi}{\partial p} = - \nabla^2 \left[\mathbf{V}_g \cdot \nabla \left(\frac{\partial \Phi}{\partial p} \right) \right] - \sigma \nabla^2 \omega \qquad (7.19)$$

where we have again assumed that σ is a constant. We next differentiate (7.16) with respect to pressure yielding

$$\frac{\partial}{\partial p} (\nabla^2 \chi) = -f_0 \frac{\partial}{\partial p} \left[\mathbf{V}_g \cdot \nabla \left(\frac{1}{f_0} \nabla^2 \Phi + f \right) \right] + f_0^2 \frac{\partial^2 \omega}{\partial p^2} \qquad (7.20)$$

Since the order of the operators on the left-hand side in (7.19) and (7.20) may be reversed, the result of subtracting (7.19) from (7.20) is to eliminate χ. After some rearrangement of terms, we obtain the *omega equation*

$$\underbrace{\left(\nabla^2 + \frac{f_0^2}{\sigma} \frac{\partial^2}{\partial p^2} \right) \omega}_{A} = \underbrace{\frac{f_0}{\sigma} \frac{\partial}{\partial p} \left[\mathbf{V}_g \cdot \nabla \left(\frac{1}{f_0} \nabla^2 \Phi + f \right) \right]}_{B} + \underbrace{\frac{1}{\sigma} \nabla^2 \left[\mathbf{V}_g \cdot \nabla \left(-\frac{\partial \Phi}{\partial p} \right) \right]}_{C} \qquad (7.21)$$

Equation (7.21) involves only derivatives in space. It is, therefore, a diagnostic equation for the field of ω in terms of the instantaneous Φ field. The omega equation, unlike the continuity equation, gives a measure of ω which does not depend on accurate observations of the horizontal wind. In fact, direct wind observations are not required at all. This method is also superior to the vorticity equation method since no knowledge of the vorticity tendency is required. In fact, only observations of Φ at a single time are required to determine the ω field using (7.21).

As was the case for the geopotential tendency equation, the terms in (7.21) are all subject to straightforward physical interpretation. The differential operator in A is identical to the operator in term A of the tendency equation (7.17). Thus, assuming that ω has a spatial distribution similar to that of χ given in (7.18) we can write

$$\left(\nabla^2 + \frac{f_0^2}{\sigma}\frac{\partial^2}{\partial p^2}\right)\omega \simeq \left[-(k^2+l^2)-\frac{1}{\sigma}\left(\frac{f_0\pi}{p_0}\right)^2\right]\omega$$

from which we can see that Term A is proportional to $-\omega$.

Term B is called the *differential vorticity advection*. Clearly this term is proportional to the rate of increase with height of the advection of absolute vorticity. To understand the role of differential vorticity advection we again consider an idealized developing baroclinic system. Figure 7.9 indicates

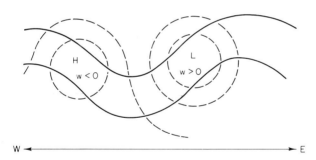

Fig. 7.9 Schematic 500-mb contours (solid lines) and 1000-mb contours (dashed lines) indicating regions of strong vertical motion due to differential vorticity advection.

schematically the geopotential contours at 500 and 1000 mb for such a system. At the centers of the surface high and surface low, designated H and L respectively, the vorticity advection at 1000 mb must be very small. However, at 500 mb the positive relative vorticity advection is a *maximum* above the surface low, while negative relative vorticity advection is strongest above the surface high. Thus, for a short-wave system where relative vorticity advection is larger than the planetary vorticity advection

$$\frac{\partial}{\partial p}[\mathbf{V}_g \cdot \nabla(\zeta_g + f)] \quad \begin{matrix} <0 \\ >0 \end{matrix} \quad \begin{matrix} \text{above point H} \\ \text{above point L} \end{matrix}$$

Recalling that $\omega \simeq -w/\rho g$ so that $\omega < 0$ implies upward vertical motion, we see from the omega equation (7.21) that differential vorticity advection is associated with rising motion above the surface low and subsidence above the surface high. This pattern of vertical motion is in fact just what is required to produce the thickness tendencies in the 1000–500-mb layer above the surface highs and lows. For example, above the surface low the positive vorticity advection creates a positive vorticity tendency. Since the geostrophic vorticity is proportional to the Laplacian of geopotential, increasing vorticity implies a falling geopotential. Thus above the surface low $\chi < 0$. Hence the 500–1000-mb thickness is decreasing in that region. Since horizontal temperature advection is small above the center of the surface low, the only way to cool the atmosphere as required by the thickness tendency is by adiabatic cooling through the vertical motion field. Thus, the vertical motion maintains a hydrostatic temperature field (that is, a field in which temperature and thickness are proportional) in the presence of differential vorticity advection. Without this compensating vertical motion either the vorticity changes at 500 mb could not remain geostrophic or the temperature changes in the 1000–500-mb layer could not remain hydrostatic.

Term C of (7.21), which is merely the negative of the horizontal Laplacian of the thickness advection, is proportional to the thickness advection

$$\nabla^2 \left[\mathbf{V}_g \cdot \nabla \left(-\frac{\partial \Phi}{\partial p} \right) \right] \propto -\mathbf{V}_g \cdot \nabla \left(-\frac{\partial \Phi}{\partial p} \right)$$

If there is warm (cold) advection, term C will be positive (negative) so that in the absence of differential vorticity advection ω would be negative (positive). Thus, as indicated in Fig. 7.10 rising motion will occur to the east of the surface low in the warm front zone, and sinking motion will occur west of the surface low behind the cold front.

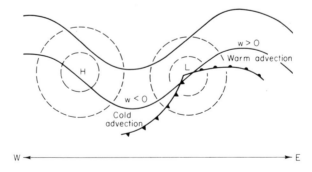

Fig. 7.10   Schematic 500-mb contours (thin solid lines), 1000 mb contours (dashed lines), and surface fronts indicating regions of strong vertical motion due to temperature advection.

Physically, this vertical motion pattern is required to keep the upper-level vorticity field geostrophic in the presence of the height changes caused by the thermal advection. For example, warm advection increases the 500–1000-mb thickness in the region of the 500-mb ridge. Thus, the geopotential height rises at the ridge and the anticyclonic vorticity must increase if geostrophic balance is to be maintained. Since vorticity advection cannot produce additional anticyclonic vorticity at the ridge, horizontal divergence is required to account for the negative vorticity tendency. Continuity of mass then requires that there be upward motion to replace the diverging air at the upper levels. By analogous arguments it can be shown that subsidence is required in the cold advection region beneath the 500-mb trough.

To summarize, we have shown as a result of scaling arguments that for synoptic scale motions where vorticity is constrained to be geostrophic and temperature is constrained to be hydrostatic, the vertical motion field is determined uniquely by the geopotential field. Further, we have shown that this vertical motion field is just that required to ensure that changes in vorticity will be geostrophic and changes in temperature will be hydrostatic. These constraints, whose importance can hardly be overemphasized, will be elaborated in Chapter 8.

7.4 Idealized Model of a Developing Baroclinic System

In the previous section, we have shown that for synoptic scale systems, the fields of vertical motion and geopotential tendency are determined to a first approximation by the three-dimensional distribution of geopotential. The results of our diagnostic analyses using the geopotential tendency and omega equations can now be combined to illustrate the essential structural characteristics of a developing baroclinic wave. For reference, we restate here the qualitative content of the tendency and omega equations:

Geopotential Tendency Equation

$$\text{Geopotential} \begin{pmatrix} \text{fall} \\ \text{rise} \end{pmatrix} \propto (\pm) \text{ Vorticity advection}$$

$$+ \text{ Rate of decrease with height of } \begin{pmatrix} \text{cold} \\ \text{warm} \end{pmatrix} \text{ advection}$$

Omega Equation

$$\begin{pmatrix} \text{Rising} \\ \text{Sinking} \end{pmatrix} \text{ motion} \propto \text{ Rate of increase with height of } (\pm) \text{ vorticity advection}$$

$$+ \begin{pmatrix} \text{warm} \\ \text{cold} \end{pmatrix} \text{ advection}$$

In Fig. 7.11 the vertical motion and horizontal divergence–convergence fields

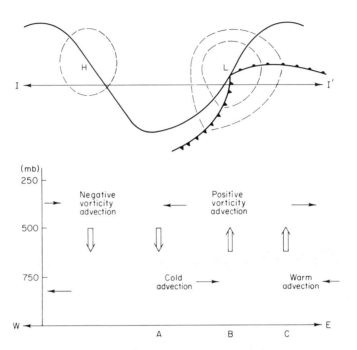

Fig. 7.11   Secondary circulation associated with a developing baroclinic wave: (top) schematic 500-mb contour (solid line), 1000-mb contours (dashed lines), and surface fronts; (bottom) vertical profile through the line II' indicating the divergent and vertical motion fields.

and their relationship to the 500 and 1000-mb geopotential fields are illustrated schematically for a developing baroclinic wave. Also indicated are the physical processes which give rise to the vertical circulation in various regions.

Additional structural features, including those which can be diagnosed with the tendency equation, are summarized in Table 7.1. In this table the signs of various physical parameters are indicated for vertical columns located at the position of (A) the 500-mb trough, (B) the surface low, and (C) the 500-mb ridge. It can be seen from this table that in all cases the vertical motion and divergence fields act to keep the temperature changes hydrostatic and vorticity changes geostrophic. Following the nomenclature of Chapter 6, we may regard the vertical and divergent motions as a *secondary circulation* imposed by the simultaneous constraints of geostrophic and hydrostatic balance. Thus, as was stated in Chapter 6, secondary circulations can be driven by processes other than friction. The secondary circulation

TABLE 7.1 *Characteristics of a Developing Baroclinic Disturbance*

Physical parameter	A 500-mb trough	B Surface low	C 500-mb ridge
$\dfrac{\partial\,\delta\Phi}{\partial t}$ (500–1000 mb)	Negative (thickness advection partly cancelled by adiabatic warming)	Negative (adiabatic cooling)	Positive (thickness advection partly canceled by adiabatic cooling)
w (500 mb)	Negative	Positive	Positive
$\dfrac{\partial\Phi}{\partial t}$ (500 mb)	Negative (differential thickness advection)	Negative (vorticity advection)	Positive differential thickness advection
$\dfrac{\partial\zeta_g}{\partial t}$ (1000 mb)	Negative (divergence)	Positive (convergence)	Positive (convergence)
$\dfrac{\partial\zeta_g}{\partial t}$ (500 mb)	Positive (convergence)	Positive (advection partly canceled by divergence)	Negative (divergence)

described in this chapter is completely independent of the circulation driven by Ekman layer pumping. In fact, it is observed that in midlatitude synoptic scale systems, the vertical velocity forced by frictional convergence in the Ekman layer is generally much smaller than the vertical velocity due to differential vorticity advection. For this reason we have neglected Ekman layer friction in developing the equations of the quasi-geostrophic system.

It is also of interest to note that the secondary circulation in a developing baroclinic system always acts to oppose the horizontal advection fields. Thus, the divergent motions tend partly to cancel the vorticity advection and the adiabatic temperature changes due to vertical motion tend to cancel partly the thermal advection. This tendency of the secondary flow to cancel partly the advective changes has important implications for the evolution of the systems which will be discussed in Chapter 8.

Finally, it should be kept in mind that the actual trajectories of air parcels moving through a synoptic system are influenced strongly by the secondary circulation. In particular, trajectory computations based on an assumption of geostrophic motion on an isobaric surface would lead to large errors in the vicinity of the warm and cold fronts as shown in Fig. 7.12. This figure further illustrates the profound difference between trajectories and streamlines for transient motion systems.

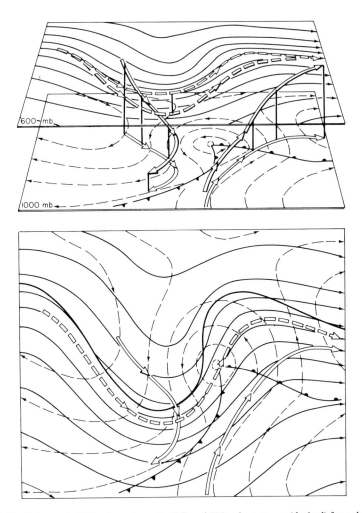

Fig. 7.12   Schematic 1000-mb contours (solid) and 600-mb contours (dashed) for a developing baroclinic wave: upper part gives perspective view indicating selected three-dimensional trajectories (arrows) and their projections on the 1000 or 600-mb surfaces. (After Palmén and Newton, 1969.)

Problems

1. Show that the static stability parameter

$$\sigma = -\frac{\alpha}{\theta}\frac{\partial \theta}{\partial p}$$

may be written in terms of Φ as

$$\sigma = \frac{\partial^2 \Phi}{\partial p^2} - \frac{1}{p}\frac{\partial \Phi}{\partial p}\left(\frac{R}{c_p} - 1\right) = \frac{1}{p^2}\left(\frac{\partial}{\partial \ln p} - \frac{R}{c_p}\right)\frac{\partial \Phi}{\partial \ln p}$$

2. Show that for an isothermal atmosphere α as defined in Problem 1 varies inversely as the square of the pressure.

3. Given the following expression for the horizontally varying part of the geopotential field

$$\Phi = cf_0[-2y(1 - p/p_0) + k^{-1}\sin k(x - ct)]$$

where $f_0 = 10^{-4}\,\text{sec}^{-1}$ is a constant Coriolis parameter, $p_0 = 1000$ mb, $k = 2 \times 10^{-8}\,\text{cm}^{-1}$, $c = 10^3\,\text{cm sec}^{-1}$, x and y are measured in centimeters, t in seconds, and p in millibars:

   (a) Obtain the horizontal divergence field corresponding to this Φ field by using the geostrophic vorticity equation. (Assume that $df/dy = 0$.)
   (b) Assuming that $\omega = 0$ at p_0, compute the ω field at 500 mb for this case by using the continuity equation.
   (c) What physical process is responsible for the presence of this vertical motion field?

4. For the geopotential distribution of Problem 3 obtain an alternative estimate of ω at 500 mb by using the hydrostatic thermodynamic energy equation. Assume that $\sigma = 10^{-4}$ in cgs units.

5. For the Φ field of Problem 3 compute ω at 500 mb using the omega equation. Assume that $\sigma = 10^{-4}$ cgs as in Problem 4. Which of these three alternative methods of computing ω should give the best results?

6. Given the following expression for the horizontally varying part of the geopotential field

$$\Phi = cf_0[-y + k^{-1}\sin(kx - \pi p/p_0)]$$

where all constants have the same values given in Problem 3, compute the geopotential tendency assuming (a) $df/dy = 0$ and (b) $df/dy = 2 \times 10^{-13}$ $\text{cm}^{-1}\,\text{sec}^{-1}$. Explain the difference in these two results. Is the *amplitude* of this disturbance changing in time?

Suggested Reference

Palmén and Newton, *Atmospheric Circulation Systems*, contains excellent descriptions of the observed structure of the mean zonal winds, planetary waves, and synoptic scale disturbances. The book is extensively illustrated with synoptic case studies which clearly illustrate the various types of atmospheric circulations.

8 | Numerical Prediction

The most important application of dynamic meteorology is in weather prediction by numerical methods. Stated simply, the objective of numerical
prediction is to predict the future state of the atmospheric circulation from
knowledge of its present state by use of the dynamical equations. To fulfill
this objective the following information is required: (1) the initial state of the
field variables, (2) a closed set of prediction equations relating the field
variables, (3) a method of integrating the equations in time to obtain the
future distribution of the field variables.

Numerical prediction is a highly specialized field which is still in a state of
rapid development. However, there do exist some fairly standard dynamical
methods of forecasting the short-term evolution of synoptic scale circulation
patterns whose general description is within the scope of this text. It is these
methods, which rely on the scaling considerations introduced in the previous
chapters, which will be emphasized in the present discussion. In particular, we
will show how the quasi-geostrophic system can be used in numerical forecasting.

8.1 Historical Background

The first attempt to numerically predict the weather was due to the
British scientist L. F. Richardson. His book, *Weather Prediction by Numerical
Process*, published in 1922, is the classical treatise in this field. In this work

Richardson showed how the differential equations governing atmospheric motions could be written approximately as a set of algebraic difference equations for values of the tendencies of various field variables at a finite number of points in space. Given the observed values of the field variables at these *grid points* the tendencies could be calculated numerically by solving the algebraic difference equations. By extrapolating the computed tendencies ahead a small increment in time an estimate of the fields at a short time in the future could be obtained. These new values of the field variables could then be used to recompute the tendencies. The new tendencies could then be used to extrapolate further ahead in time, etc. Even for short-range forecasting over a small area of the earth, this procedure requires an enormous number of arithmetic operations. Richardson did not foresee the development of high-speed digital computers. He estimated that a work force of 64,000 people would be required just to keep up with the weather on a global basis.

Despite the tedious labor involved, Richardson worked out one example forecast for surface pressure tendencies at two grid points. Unfortunately, the results were very unimpressive. Predicted pressure changes were an order of magnitude larger than those observed. At the time this failure was thought to be due primarily to the poor initial data available—especially the absence of upper-air soundings. However, it is now known that there were other even more serious problems with Richardson's scheme.

After Richardson's failure to obtain a reasonable forecast, numerical prediction was not again attempted for many years. Finally, after World War II interest in numerical prediction revived due partly to the vast expansion of the meteorological observation network, which provided much improved initial data, but even more importantly to the development of digital computers which made the enormous volume of arithmetic operations required in a numerical forecast feasible. At the same time it was realized that Richardson's scheme was not the simplest possible scheme for numerical prediction. Richardson's equations governed not only the slow-moving meterologically important motions, but also included high-speed sound and gravity waves as solutions. These high-speed sound and gravity waves are in nature very weak in amplitude. However, for reasons that will be explained later, if Richardson had carried his numerical calculation beyond the initial time step, these oscillations would have amplified spuriously thereby introducing so much "noise" in the solution that the meteorologically relevant disturbances would have been obscured.

J. G. Charney in 1948 showed how the dynamical equations could be simplified by systematic introduction of the geostrophic and hydrostatic assumptions so that the sound and gravity oscillations were filtered out. The equations which resulted from Charney's filtering approximations were essentially those of the quasi-geostrophic model. A special case of this model, the so-called equivalent barotropic model, was used in 1950 to make the first numerical forecast.

The first model provided forecasts only of the geopotential, and hence the geostrophic winds, near 500 mb. Thus, this model did not forecast "weather" in the usual sense, but only the winds at 500 mb. Later multilevel models based on the quasi-geostrophic theory could, however, be used to diagnose the large-scale vertical motion field at the forecast time. Since vertical motion is correlated with precipitation, this information could be used by forecasters as an aid in predicting the local weather associated with large-scale circulations.

With the development of vastly more powerful computers and more sophisticated modeling techniques the emphasis in numerical forecasting has now returned to models which are quite similar to Richardson's formulation.

8.2   Filtering of Sound and Gravity Waves

One difficulty in directly applying the unsimplified equations of motion is that meteorologically important motions are easily lost in the noise introduced by large-amplitude sound and gravity waves which may arise as a result of errors in the initial data, and then spuriously amplify by a process called *computational instability*. As an example of how this problem might arise, we know that on the synoptic scale the pressure and density fields are in hydrostatic balance to a very good approximation. As a consequence. vertical accelerations are extremely small. However, if the pressure and density fields were determined independently by observations, as would be the case if the complete equations were used, small errors in the observed fields would lead to large errors in the computed vertical acceleration simply because the vertical acceleration is the very small difference between two large forces—the vertical pressure gradient and gravity. Such spurious accelerations would appear in the computed solution as high-speed sound waves of very large amplitude. In a similar fashion errors in the initial velocity and pressure fields would lead to spuriously large horizontal accelerations since the horizontal acceleration results from the small difference between the Coriolis and pressure gradient forces. Such spurious horizontal accelerations would excite both sound and gravity waves.

In order to overcome this problem, the simplest procedure is to simplify the governing equations to remove the physical mechanisms responsible for the occurrence of the unwanted oscillations, while still preserving the meteorologically important motions. To see how such a "filtering" of the equations can be accomplished we next consider the qualitative nature of sound and gravity waves.

Sound waves, sometimes called acoustic waves, are longitudinal waves, that

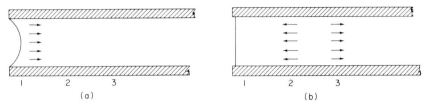

Fig. 8.1 Schematic diagram illustrating the propagation of a sound wave in a tube with a flexible diaphragm at the left end.

is, waves in which the particle oscillations are parallel to the direction of propagation. Sound is propagated by the alternating adiabatic compression and expansion of the medium. As an example, in Fig. 8.1 we show a schematic section along a tube which has a diaphragm at its left end. If the diaphragm is deflected to the right by striking it as shown in Fig. 8.1a, the air between points 1 and 2 will be compressed. Thus, there will be a pressure gradient which will accelerate the air into the region between points 2 and 3. As a result the air will be compressed between points 2 and 3, and as shown in Fig. 8.1b the pressure gradient force will cause the air to accelerate out of that region, thus creating a new region of compression to the right of point 3. By this continual process of adiabatic increase and decrease of the pressure through alternating compression and rarefaction, the disturbance excited by striking the membrane will move rightward down the pipe. Individual air parcels do not, however, have a net rightward motion, they only oscillate back and forth while the pressure pattern moves rightward at the speed of sound.

If the pipe in Fig. 8.1 is tipped up in the vertical, it can be used to generate vertically traveling sound waves. If we now require that pressure be hydrostatic, the pressure at any point along the pipe must be determined solely by the *weight* of the air above that point. Hence, the vertical pressure gradient cannot be influenced by adiabatic compression. Therefore, sound waves cannot propagate *vertically* if it is assumed that the motions are hydrostatic. Replacement of the vertical momentum equation by the hydrostatic approximation is thus sufficient to filter out ordinary sound waves. However, a hydrostatically balanced atmosphere can still support a special class of *horizontally* propagating acoustic waves. In this type of acoustic wave the vertical velocity is zero (neglecting orographic effects), but the pressure oscillates at the lower boundary. (These waves are discussed further in Section 9.3.) To filter out this type of oscillation it is necessary to require that $\omega \equiv dp/dt = 0$ at the lower boundary. This boundary condition is most easily applied by formulating the equations in isobaric coordinates. In that case the condition $\omega = 0$ at the lower boundary is a natural first approximation for geostrophically scaled motions as can be seen from the discussion in Section 4.4.

Thus, as a result of the minimum simplification necessary to filter out sound waves, the prediction equations become the isobaric horizontal momentum equation, continuity equation, and hydrostatic thermodynamic energy equation subject to the condition $\omega = 0$ at the lower boundary. These may be written using vector notation as

$$\left(\frac{\partial}{\partial t} + \mathbf{V} \cdot \nabla\right)\mathbf{V} + \omega \frac{\partial \mathbf{V}}{\partial p} + f\mathbf{k} \times \mathbf{V} = -\nabla\Phi \qquad (8.1)$$

$$\mathbf{V} \cdot \mathbf{V} + \frac{\partial \omega}{\partial p} = 0 \qquad (8.2)$$

$$\left(\frac{\partial}{\partial t} + \mathbf{V} \cdot \nabla\right)\left(\frac{\partial \Phi}{\partial p}\right) + \sigma\omega = 0 \qquad (8.3)$$

where as before $\sigma \equiv -(\alpha/\theta)(\partial\theta/\partial p)$.

The system (8.1)–(8.3) together with the condition $\omega = 0$ at the lower boundary no longer contains the mechanism for sound-wave transmission but is still capable of describing *gravity waves*. Gravity waves are waves in which buoyancy acts as the restoring force on parcels displaced vertically. The disturbance which propagates across the surface of a pond when a stone is tossed in the water is a simple example of a gravity wave. Similar waves can occur in the atmosphere provided that the static stability is positive so that a parcel displaced vertically will tend to oscillate about its original position.

The mechanism for propagation of gravity waves is most easily understood by considering a disturbance on the free surface of an incompressible fluid such as water. Suppose, as shown in Fig. 8.2, there is a depression in the free surface centered at the origin $x = 0$ at time $t = t_0$. As a result of this disturbance there will be a horizontal pressure gradient which will cause an acceleration toward the origin. Thus, fluid will converge horizontally into the column of water centered at $x = 0$. However, since the fluid is incompressible, this convergence must be compensated by horizontal divergence on both sides. Thus at time $t = t_1$ there will be depressions on both sides of the original disturbance. Again, the unbalanced horizontal pressure gradients will cause accelerations into the depressions and the result will be a continual outward propagation of the disturbance due to alternating horizontal convergence and divergence in individual columns of fluid.

From the above example it should be apparent that a divergent horizontal velocity field which changes in time is essential for the propagation of gravity waves. In fact it turns out that neglecting thg local rate of change of the horizontal divergence in computing the balance between the mass and

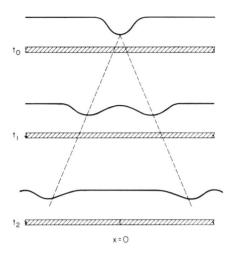

Fig. 8.2 The propagation of a surface gravity wave initiated by a depression at time t_0.

velocity fields is sufficient to filter out the time-dependent gravity waves. Of course, the momentum equation (8.1) does not explicitly contain the local derivative of $\mathbf{V} \cdot \mathbf{V}$. However, the prediction system can be rewritten so that this term does appear if we replace the horizontal equations of motion by the vorticity and divergence equations which are both differentiated forms of the momentum equation. Although we have previously discussed the vorticity equation in Sections 5.4.2. and 7.3, the derivation will be briefly repeated here in vectorial form for completeness.

Using the vector identity

$$(\mathbf{V} \cdot \mathbf{V})\mathbf{V} = \mathbf{V}\left(\frac{\mathbf{V} \cdot \mathbf{V}}{2}\right) + \mathbf{k} \times \mathbf{V}\zeta$$

where $\zeta = \mathbf{k} \cdot \mathbf{V} \times \mathbf{V}$ is as before the vertical component of vorticity, we can rewrite (8.1) in the form

$$\frac{\partial \mathbf{V}}{\partial t} = -\mathbf{V}\left(\Phi + \frac{\mathbf{V} \cdot \mathbf{V}}{2}\right) - \mathbf{k} \times \mathbf{V}(\zeta + f) - \omega \frac{\partial \mathbf{V}}{\partial p} \tag{8.4}$$

The vorticity and divergence equations are now obtained by operating on (8.4) with the vector operators $\mathbf{k} \cdot \mathbf{V} \times (\ )$ and $\mathbf{V} \cdot (\ )$, respectively. The resulting equations are the vorticity equation,

$$\frac{\partial \zeta}{\partial t} = -\mathbf{V} \cdot \mathbf{V}(\zeta + f) - \omega \frac{\partial \zeta}{\partial p} - (\zeta + f)\mathbf{V} \cdot \mathbf{V} + \mathbf{k} \cdot \left(\frac{\partial \mathbf{V}}{\partial p} \times \mathbf{V}\omega\right) \tag{8.5}$$

and the divergence equation,

$$\frac{\partial}{\partial t}(\mathbf{V}\cdot\mathbf{V}) = -\nabla^2\left(\Phi + \frac{\mathbf{V}\cdot\mathbf{V}}{2}\right) - \nabla\cdot[\mathbf{k}\times\mathbf{V}(\zeta + f)]$$

$$- \omega\frac{\partial}{\partial p}(\mathbf{V}\cdot\mathbf{V}) - \frac{\partial\mathbf{V}}{\partial p}\cdot\nabla\omega \quad (8.6)$$

Equations (8.5)–(8.6) are independent scalar equations which can be used in place of the horizontal equations of motion in our prediction system. If we now simply set the term on the left-hand side in (8.6) equal to zero, we will eliminate all solutions corresponding to time dependent gravity waves. This is the minimum simplification required to filter gravity waves.

8.3 Filtered Forecast Equations

Merely neglecting the term on the left-hand side turns (8.6) into a rather intractable diagnostic relationship between ω, $\mathbf{V}$, and Φ. However, further simplifications can be made with the aid of scale analysis. Indeed, for synoptic scale motions the neglect of $\partial(\mathbf{V}\cdot\mathbf{V})/\partial t$ in (8.6) can itself be justified on the basis of scaling arguments. Thus, neglecting second-order terms in (8.6) is sufficient to filter the gravity waves.

In order to exhibit the connection between scaling and filtering more clearly it is convenient to partition the horizontal velocity field into nondivergent and divergent components. It is shown in Appendix C that any velocity field can be divided into a *nondivergent* part $\mathbf{V}_\psi$ plus a *divergent* part $\mathbf{V}_e$ such that

$$\mathbf{V} = \mathbf{V}_\psi + \mathbf{V}_e \quad (8.7)$$

where

$$\nabla\cdot\mathbf{V}_\psi = 0 \quad \text{and} \quad \nabla\times\mathbf{V}_e = 0$$

If the velocity field is two dimensional, the nondivergent part can be expressed in terms of the *streamfunction* ψ defined by letting

$$\mathbf{V}_\psi = \mathbf{k}\times\nabla\psi \quad (8.8)$$

or in cartesian components,

$$u_\psi = -\frac{\partial\psi}{\partial y}, \qquad v_\psi = \frac{\partial\psi}{\partial x}$$

from which it is easily verified that

$$\nabla\cdot\mathbf{V}_\psi = 0 \quad \text{and} \quad \zeta \equiv \mathbf{k}\cdot\nabla\times\mathbf{V} = \nabla^2\psi$$

Because the isolines of ψ correspond to streamlines for the nondivergent velocity and the distance separating the isolines of ψ is proportional to the magnitude of the nondivergent velocity, the spatial distribution of $\mathbf{V}_\psi$ can be easily pictured by plotting lines of constant ψ on a map.

In our scale analysis of the vorticity equation in Chapter 5 we showed that for midlatitude synoptic scale motions $\mathbf{V}$ must be quasi-nondivergent, that is,

$$|\mathbf{V}_\psi| \gg |\mathbf{V}_e|$$

Thus to a first approximation we can replace $\mathbf{V}$ by $\mathbf{V}_\psi$ everywhere in (8.5) and (8.6) *except* in terms involving the horizontal divergence. It is left as a problem for the reader to show that the resulting approximate vorticity and divergence equations, valid for midlatitude synoptic scale motions, become

$$\frac{\partial \zeta}{\partial t} = -\mathbf{V}_\psi \cdot \nabla(\zeta + f) - f_0 \nabla \cdot \mathbf{V}_e \tag{8.9}$$

and

$$\nabla^2 \Phi = -f_0 \nabla \cdot (\mathbf{k} \times \mathbf{V}_\psi) = f_0 \nabla^2 \psi \tag{8.10}$$

We note that as a result of the scaling approximations all terms involving the divergence have been eliminated from (8.6) so that the gravity-wave solutions have been filtered. Equation (8.10) simply states that to a first approximation the vorticity is the vorticity of the geostrophic wind computed using a constant midlatitude value of the Coriolis parameter. Thus, for midlatitude synoptic scale motions the streamfunction is given approximately by the relation $\psi \simeq \Phi/f_0$. Hence, the geopotential field on a constant pressure chart is approximately proportional to the streamfunction field, and to the same order of approximation

$$\mathbf{V}_\psi \simeq \frac{\mathbf{k} \times \nabla \Phi}{f_0} \tag{8.11}$$

The geostrophic vorticity equation and hydrostatic thermodynamic energy equation can now be written in terms of ψ and ω as

$$\frac{\partial}{\partial t} \nabla^2 \psi = -\mathbf{V}_\psi \cdot \nabla(\nabla^2 \psi + f) + f_0 \frac{\partial \omega}{\partial p} \tag{8.12}$$

$$\frac{\partial}{\partial t}\left(\frac{\partial \psi}{\partial p}\right) = -\mathbf{V}_\psi \cdot \nabla\left(\frac{\partial \psi}{\partial p}\right) - \frac{\sigma}{f_0} \omega \tag{8.13}$$

Aside from the change of notation this set of equations is identical to the quasi-geostrophic system derived in Chapter 7.

Differentiation of (8.13) with respect to p after multiplying through by

f_0^2/σ and adding the result to (8.12) gives the quasi-geostrophic *potential vorticity equation.*

$$\left(\frac{\partial}{\partial t} + \mathbf{V}_\psi \cdot \mathbf{V}\right)q = 0 \tag{8.14}$$

where

$$q = \mathbf{V}^2\psi + f + f_0^2 \frac{\partial}{\partial p}\left(\frac{1}{\sigma}\frac{\partial\psi}{\partial p}\right)$$

and we have again assumed that σ is a function of pressure only. This equation, which is equivalent to the diagnostic geopotential tendency equation (7.17), states that the geostrophic potential vorticity q is *conserved following the nondivergent wind* in pressure coordinates. This conservation law is the basis for most of the numerical prediction schemes discussed in this chapter. Comparing with Eq. (5.12) we see that this law is similar to, but not identical to, the more general potential vorticity conservation law of Chapter 5.

Again, as in Chapter 7, if the time derivatives are eliminated between (8.12) and (8.13), we obtain the diagnostic omega equation

$$\left(\mathbf{V}^2 + \frac{f_0^2}{\sigma}\frac{\partial^2}{\partial p^2}\right)\omega = \frac{f_0}{\sigma}\frac{\partial}{\partial p}[\mathbf{V}_\psi \cdot \mathbf{V}(\mathbf{V}^2\psi + f)] - \frac{f_0}{\sigma}\mathbf{V}^2\left[\mathbf{V}_\psi \cdot \mathbf{V}\left(\frac{\partial\psi}{\partial p}\right)\right] \tag{8.15}$$

Equation (8.15) can be used to diagnose the ω field at any instant provided that the ψ field is known.

Now since according to (8.14) q is conserved following the nondivergent wind, prediction with the quasi-geostrophic system in principle merely involves advecting the q field with the nondivergent wind at each point. Of course the nondivergent wind itself changes in time as q is advected along. Thus, after extrapolating ahead for a short time, we must invert the operator

$$\mathbf{V}^2 + f_0^2 \frac{\partial}{\partial p}\left(\frac{1}{\sigma}\frac{\partial}{\partial p}\right)$$

to obtain ψ from the computed field of q. We may then use the new ψ field to recompute the nondivergent wind and advect q along for another short time. By continually repeating this process a forecast for the distribution of ψ at any time in the future can be prepared.

In practice it has not been possible to solve a forecast equation such as (8.14) for all standard data levels in the vertical. Too much machine time and storage would be required. Instead, models have been developed which simplify the vertical structure by referring to one, or at most a few, layers. Following the historical development of the subject, we will discuss the simplest models first.

8.4 One-Parameter Models

The simplest type of model is one which refers to only one data level in the vertical. Such a model is called a one-parameter model. One-parameter models are inherently barotropic for the inclusion of thickness advection, which is crucial to baroclinic processes, requires knowledge of Φ at more than a single level.

8.4.1 THE BAROTROPIC MODEL

The simplest one-parameter model is obtained by assuming that there exists a level of nondivergence in the atmosphere. This level is usually taken to be the 500-mb surface. Thus, we assume that

$$\mathbf{V} \cdot \mathbf{V} = -\frac{\partial \omega}{\partial p} = 0 \qquad \text{at 500 mb.}$$

This assumption, although certainly not true everywhere, is partly justified observationally since computed vertical motion patterns generally show a maximum in ω near 500 mb. Assuming that 500 mb is a level of nondivergence, the vorticity equation (8.9) simplifies to

$$\frac{\partial}{\partial t}\mathbf{V}^2\psi = -\mathbf{V}_\psi \cdot \mathbf{V}(\mathbf{V}^2\psi + f) \tag{8.16}$$

This equation allows one to compute the evolution of the flow at the level of nondivergence only. No vertical coupling is involved. Thus, no predictions for levels above or below the nondivergent level are possible. The barotropic vorticity equation (8.16) is an exact model only for a homogeneous incompressible fluid confined between rigid, frictionless horizontal boundaries. Clearly the atmosphere does not meet these basic requirements. However, because midtropospheric synoptic scale flow is in general quasi-nondivergent, (8.16) is a useful approximate forecast equation. Furthermore, unlike the potential vorticity equation (8.14), the barotropic vorticity equation does not require that the motion be quasi-geostrophic. It is only necessary that the horizontal velocity be quasi-nondivergent. Thus, as we shall see in Chapter 12, (8.16) remains a useful approximation even in equatorial regions. In fact, outside areas of active precipitation, (8.16) tends to be even a better approximation in the tropics than it is at middle latitudes.

8.4.2 THE EQUIVALENT BAROTROPIC MODEL

A slightly improved forecasting model can be obtained by averaging in the vertical so that some account is taken of the variation of wind with height yet only one level is still required to solve the equation. The main simplifying assumption required for a vertically averaged model is based on the observed tendency for isotherms and isobars to be nearly parallel above the friction layer—at least in the absence of pronounced development of systems. In the equivalent barotropic model it is assumed that the isotherms and isobars are exactly parallel. The thermal wind is then in the same direction as the geostrophic wind at all heights. Thus, the geostrophic wind will vary in magnitude with height but not in direction.

With this assumption about the vertical structure, the nondivergent wind can be written

$$\mathbf{V}_\psi(x, y, p) = A(p)\langle\mathbf{V}_\psi(x, y)\rangle \qquad (8.17)$$

The angle brackets here denote a vertical average

$$\langle\ (\ )\ \rangle = \frac{1}{p_0} \int_0^{p_0} (\ )\, dp \qquad (8.18)$$

where $p_0 = 1000$ mb. Thus, $A(p)$ is a weighting function depending only on pressure which gives the vertical dependence of the magnitude of the nondivergent wind. From the definitions of ζ and ψ we also can write

$$\zeta = A(p)\langle\zeta\rangle, \qquad \psi = A(p)\langle\psi\rangle$$

Using the above notation, the vorticity equation (8.12) can be rewritten as

$$\frac{\partial}{\partial t}[A(p)\,\nabla^2\langle\psi\rangle] = -A(p)^2\langle\mathbf{V}_\psi\rangle \cdot \nabla(\nabla^2\langle\psi\rangle) - A(p)\frac{\partial\langle\psi\rangle}{\partial x}\,\beta + f_0\frac{\partial\omega}{\partial p}$$
$$(8.19)$$

We next average (8.19) in the vertical by applying (8.18), noting that

$$\langle A(p)\rangle = \frac{1}{p_0}\int_0^{p_0} A(p)\, dp = 1$$

and $\omega(0) = 0$. The result is

$$\frac{\partial}{\partial t}\nabla^2\langle\psi\rangle = -\langle A(p)^2\rangle\langle\mathbf{V}_\psi\rangle \cdot \nabla(\nabla^2\langle\psi\rangle) - \beta\frac{\partial\langle\psi\rangle}{\partial x} + \frac{f_0\,\omega(p_0)}{p_0} \qquad (8.20)$$

It is important to observe that although $\langle A(p)\rangle = 1, \langle A(p)^2\rangle$ is in general different from unity:

$$\langle A(p)^2\rangle = \frac{1}{p_0}\int_0^{p_0} A(p)^2\, dp$$

We now define a level p^*, called the starred level, according to

$$\mathbf{V}_\psi(x, y, p^*) \equiv \mathbf{V}^* = \langle A(p)^2\rangle\langle\mathbf{V}_\psi\rangle \qquad (8.21)$$

so that

$$\psi(x, y, p^*) \equiv \psi^* = \langle A(p)^2 \rangle \langle \psi \rangle$$

Multiplying through by $\langle A(p)^2 \rangle$ in (8.20) and using the definitions in (8.21) we find that at the starred level the vorticity equation is

$$\frac{\partial}{\partial t} \nabla^2 \psi^* = -\mathbf{V}^* \cdot \nabla(\nabla^2 \psi^* + f) + \frac{\langle A(p)^2 \rangle f_0\, \omega(p_0)}{p_0} \qquad (8.22)$$

Therefore, at the starred level the prediction equation reduces to the barotropic vorticity equation with one additional term due to vertical motion at the lower boundary.

The problem in using this model, aside from questions as to the validity of the basic assumption concerning the vertical variation of the wind, is that we must determine the starred level. In practice this is not done for each forecast, but rather a climatologically derived starred level computed using seasonal mean data is generally used. Since in the mean the zonal velocity is much larger than the meridional velocity, the profile of $u(p)$ is usually used to compute the starred level.

Determination of the starred level is illustrated in Fig. 8.3. Averaging $u(p)$ in the vertical gives $A(p) = u(p)/\langle u \rangle$ from which $A(p)^2$ and $\langle A(p)^2 \rangle$ are easily computed. The starred level is just that level at which $A(p) = \langle A(p)^2 \rangle$.

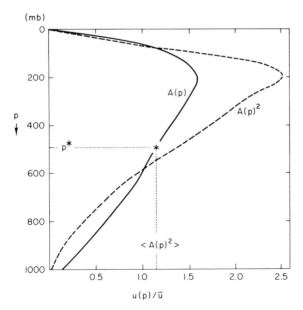

Fig. 8.3   Method of computing the equivalent barotropic level. See text for details.

Actually there exist two starred levels, one near 500 mb and one near 100 mb. However, it is generally the midtropospheric level that is of interest.

Equation (8.22) still cannot be used alone for prediction because it is a single equation in two variables, ψ^* and ω_0. Thus, it is desirable to express ω_0 in terms of ψ^*. We recall that by definition

$$\omega \equiv \frac{dp}{dt} = \frac{\partial p}{\partial t} + \mathbf{V} \cdot \nabla p + w \frac{\partial p}{\partial z} \tag{8.23}$$

so that as shown in Section 4.3

$$\omega \simeq - \rho g w$$

Therefore, at the lower boundary

$$\omega(p_0) \simeq - \rho_0\, g w(z_0) \tag{8.24}$$

where $w(z_0)$ is the vertical velocity at the height z_0 corresponding to pressure p_0. As a first approximation we assume that z_0 is coincident with the ground.[1] Then $w(z_0)$ is zero over flat land, but over sloping terrain where the flow near the ground is deflected vertically by the sloping surface

$$w(z_0) \simeq \frac{dh}{dt} \simeq \mathbf{V}_\psi(z_0) \cdot \nabla h$$

where h is the ground height, and $\mathbf{V}_\psi(z_0)$ is the horizontal velocity at z_0. The equalities in the above expression are only approximate because it must be remembered that we are applying the boundary condition at the height z_0 of the 1000-mb surface which of course is not coincident with the actual surface at $z = h$.

From our previous assumption for the vertical variation of $\mathbf{V}_\psi$ (8.17) and (8.21) we can write

$$\mathbf{V}_\psi(z_0) = A(p_0)\langle \mathbf{V}_\psi \rangle = \frac{A(p_0)}{\langle A(p)^2 \rangle}\, \mathbf{V}^*$$

so that

$$w(z_0) \simeq \frac{A(p_0)}{\langle A(p)^2 \rangle}\, \mathbf{V}^* \cdot \nabla h \tag{8.25}$$

Using (8.24) and (8.25) to eliminate $\omega(p_0)$ in the vorticity equation for the

---

[1] Actually z_0 should be taken as the top of the Ekman boundary layer since the present inviscid model obviously does not apply in the boundary layer. In that case, as shown in Section 6.3, $w(z_0)$ would be proportional to $\zeta(z_0)$. However, for simplicity we will neglect the friction layer in the present discussion.

starred level (8.22), we obtain the equivalent barotropic model

$$\frac{\partial}{\partial t} \nabla^2 \psi^* + \mathbf{V}^* \cdot \mathbf{V} \left[\nabla^2 \psi^* + f + \frac{A(p_0) f_0 g h}{R T_0} \right] = 0 \qquad (8.26)$$

where we have used the equation of state to eliminate ρ_0 in the last term.

Assuming that the topography $h(x, y)$ is known (8.26) can be integrated in time to give the evolution of ψ^*. To the extent that the assumption concerning the variation of ψ with pressure is correct, the results can be used to deduce the evolution of ψ at other levels. Formally, (8.26) is equivalent to the barotropic model (8.16) except that (8.26) includes orography in a crude form, and does not require the assumption of a nondivergent level. For these reasons, nearly all operational barotropic prognoses are based on the equivalent barotropic model.

8.5 A Two-Parameter Baroclinic Model

In Chapter 7 we showed that thermal advection processes are essential to the development of synoptic systems. The barotropic and equivalent barotropic models do not allow temperature advection, therefore they cannot forecast the development of new systems. In fact barotropic models are really merely extrapolation formulas which state that the vertical vorticity distribution at any instant is advected isobarically by the windfield. The fact that barotropic prognoses are quite effective in predicting the evolution of midtropospheric flow for periods of up to two or three days indicates that in the short range, barotropic vorticity advection is the primary mechanism governing the flow. This fact simply reflects the quasi-horizontal and quasi-nondivergent character of midlatitude synoptic scale flows.

However, mere advection of the initial circulation field is clearly not satisfactory if we wish to produce forecasts which are consistently reliable. It is necessary, in addition, to predict the development of new systems. To include thermal advection processes which are essential for baroclinic development we must use a model which involves more than one data level in the atmosphere. We must also explicitly use the thermodynamic energy equation. More than a single data level is required because to compute temperature advection we must know the thickness which in turn requires measurement of the difference in geopotential between two levels in the vertical.

The simplest model which can incorporate the baroclinic temperature advection process is one in which the geopotential is predicted at *two* levels. The thickness or mean temperature can then be represented as the difference in geopotential between those two levels. To derive this model we divide the atmosphere into four discrete layers bounded by surfaces numbered 0 to 4 as shown in Fig. 8.4.

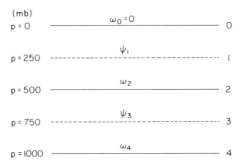

Fig. 8.4 Arrangement of variables in the vertical for the two-parameter baroclinic model.

We now apply the vorticity equation (8.12) at the two levels designated as 1 and 3. To do this we must evaluate the divergence term $\partial\omega/\partial p$ at each level. Using *finite difference* approximations to the vertical derivatives we obtain

$$\left(\frac{\partial\omega}{\partial p}\right)_1 \simeq \frac{\omega_2 - \omega_0}{\Delta p}, \qquad \left(\frac{\partial\omega}{\partial p}\right)_3 \simeq \frac{\omega_4 - \omega_2}{\Delta p}$$

where Δp is the pressure interval between levels 0–2 and 2–4, and subscript notation is used to designate the vertical level (0–4) for each dependent variable. The resulting vorticity equations are

$$\frac{\partial}{\partial t}\nabla^2\psi_1 = -(\mathbf{k}\times\nabla\psi_1)\cdot\nabla(\nabla^2\psi_1 + f) + \frac{f_0}{\Delta p}\omega_2 \tag{8.27}$$

$$\frac{\partial}{\partial t}\nabla^2\psi_3 = -(\mathbf{k}\times\nabla\psi_3)\cdot\nabla(\nabla^2\psi_3 + f) - \frac{f_0}{\Delta p}\omega_2 \tag{8.28}$$

where for simplicity we have assumed that $\omega_4 = 0$, which will be approximately true for flat ground.

We next write the thermodynamic energy equation (8.13) for level 2. Here we evaluate $\partial\psi/\partial p$ using the difference formula

$$\left(\frac{\partial\psi}{\partial p}\right)_2 \simeq \frac{\psi_3 - \psi_1}{\Delta p}$$

The result is

$$\frac{\partial}{\partial t}(\psi_1 - \psi_3) = -(\mathbf{k}\times\nabla\psi_2)\cdot\nabla(\psi_1 - \psi_3) + \frac{\sigma\,\Delta p}{f_0}\omega_2 \tag{8.29}$$

The first term on the right-hand side in (8.29) is the advection of the 250–750-mb thickness by the wind at 500 mb, which is taken to be the mean wind in the layer. However ψ_2, the 500-mb streamfunction, is not a predicted quantity in this model. Therefore, ψ_2 must be obtained by linearly interpolating between 250 and 750 mb:

$$\psi_2 = \tfrac{1}{2}(\psi_1 + \psi_3) \tag{8.30}$$

If this interpolation formula is used, (8.27)–(8.29) become a closed set of prediction equations in the variables ψ_1, ψ_2, and ψ_3.

We define the dimensionless parameter

$$\lambda^2 \equiv \frac{f_0^{\,2}}{\sigma(\Delta p)^2}$$

We now eliminate ω_2 between (8.27)–(8.29) to obtain two equations in ψ_1 and ψ_3 alone. We first simply add (8.27) and (8.28) to obtain

$$\frac{\partial}{\partial t}\nabla^2(\psi_1 + \psi_3) = -(\mathbf{k} \times \nabla\psi_1)\cdot\nabla(\nabla^2\psi_1 + f) - (\mathbf{k} \times \nabla\psi_3)\cdot\nabla(\nabla^2\psi_3 + f)$$

$$\tag{8.31}$$

We next subtract (8.28) from (8.27) and add the result to $-2\lambda^2$ times (8.29) to get

$$\begin{aligned}
\frac{\partial}{\partial t}[(\nabla^2 - 2\lambda^2)(\psi_1 - \psi_3)] = &-(\mathbf{k} \times \nabla\psi_1)\cdot\nabla(\nabla^2\psi_1 + f) \\
&+ (\mathbf{k} \times \nabla\psi_3)\cdot\nabla(\nabla^2\psi_3 + f) \\
&+ 2\lambda^2(\mathbf{k} \times \nabla\psi_2)\cdot\nabla(\psi_1 - \psi_3) \quad (8.32)
\end{aligned}$$

Essentially, (8.31) states that the local rate of change of the vertically averaged vorticity (that is, the average of the 250- and 750-mb vorticities) is equal to the average of the 250- and 750-mb vorticity advections. Thus, (8.31) governs the barotropic part of the flow.

On the other hand, (8.32) may be regarded as an equation for the thickness tendency. Equation (8.32) states that the local rate of change of the 250–750-mb thickness is proportional to the difference between the vorticity advections at 250 and 750 mb plus the thermal advection. Thus, the physical mechanisms expressed by the terms on the right-hand side in (8.32) are identical to those contained in the diagnostic tendency equation (7.17).

We can also get an omega equation for the two-parameter model by combining (8.27)–(8.29) so that the time derivatives are eliminated. It is merely necessary to operate on (8.29) with ∇^2 then add (8.28) and subtract (8.27) from the result. After rearranging terms we get

$$\begin{aligned}
(\nabla^2 - 2\lambda^2)\omega_2 = \frac{f_0}{\sigma\Delta p}\,\{\nabla^2[(\mathbf{k} \times \nabla\psi_2)\cdot\nabla(\psi_1 - \psi_3)] \\
+ (\mathbf{k} \times \nabla\psi_3)\cdot\nabla(\nabla^2\psi_3 + f) \\
- (\mathbf{k} \times \nabla\psi_1)\cdot\nabla(\nabla^2\psi_1 + f)\} \quad (8.33)
\end{aligned}$$

Equation (8.33) is simply a two-layer finite-difference approximation to the omega equation (8.15) previously discussed. In fact (8.33) could have been obtained directly from (8.15) by approximating the vertical derivatives using finite differences centered at 500 mb.

Since the two-parameter model has been derived from the equations of the quasi-geostrophic system, the physical mechanisms described in the model should be those which were discussed in Chapter 7. In the remainder of this section we will verify that this is indeed the case. We will also show that the simplicity of the two-parameter model helps elucidate certain aspects of the divergent secondary circulations not previously discussed.

To review briefly the properties of the quasi-geostrophic system, we recall first that this system requires that the atmosphere simultaneously satisfy two constraints: (1) vorticity changes are geostrophic, and (2) temperature changes are hydrostatic. Thus both vorticity and temperature are proportional to derivatives of the geopotential field. In order that these constraints both be satisfied the omega field must at every instant be adjusted so that the divergent motions keep the vorticity changes geostrophic and the vertical motions keep the temperature changes hydrostatic.

These properties of the quasi-geostrophic system are clearly revealed in the two-parameter model by examining the individual terms of the omega equation (8.33). We previously showed that for a wave-type disturbance the horizontal Laplacian operator can be approximated by writing

$$\nabla^2 \simeq -(k^2 + l^2)$$

where $k = 2\pi/L_x$, $l = 2/L_y$ with L_x and L_y the wavelengths of the disturbance in the x and y directions, respectively. Thus we can write (8.33) approximately as

$$\omega_2 \simeq \frac{f_0}{\sigma \Delta p (k^2 + l^2 + 2\lambda^2)} \{(k^2 + l^2)[(\mathbf{k} \times \nabla\psi_2) \cdot \nabla(\psi_1 - \psi_3)]$$
$$+ (\mathbf{k} \times \nabla\psi_1) \cdot \nabla(\nabla^2\psi_1 + f) - (\mathbf{k} \times \nabla\psi_3) \cdot \nabla(\nabla^2\psi_3 + f)\} \qquad (8.34)$$

The first term on the right-hand side is proportional to minus the thickness advection by the 500-mb nondivergent wind. Thus, for example, in the case of cold advection where $\mathbf{k} \times \nabla\psi_2$ would have a component positive parallel to $\nabla(\psi_1 - \psi_3)$ this term would tend to make ω_2 positive. For warm advection, on the other hand, the opposite signs would apply. Therefore, we may summarize as follows:

$$\begin{pmatrix} \text{cold} \\ \text{warm} \end{pmatrix} \text{advection implies} \begin{pmatrix} \text{sinking} \\ \text{rising} \end{pmatrix} \text{motion at 500 mb}$$

The divergent motions associated with this vertical motion field are indicated in Fig. 8.5 for a developing wave in the two-layer model. Linear

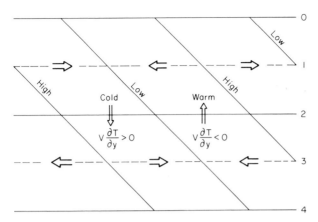

Fig. 8.5 Secondary circulation due to north–south thermal advection in the two-parameter model.

interpolation has been used between levels so that the trough and ridge axes are straight lines tilted back toward the west with height. The divergence field associated with the subsiding branch of the vertical circulation contributes a positive vorticity tendency at the 250-mb trough and a negative vorticity tendency at 750 mb. Conversely, in the region of ascending motion the divergence field contributes a negative vorticity tendency at the 250-mb ridge and a positive tendency at 750 mb. Since in all cases these vorticity tendencies tend to increase the extreme values of vorticity at the troughs and ridges, this secondary circulation system will act to increase the strength of the disturbances. The disturbances will amplify most rapidly when the slope of the trough and ridge axes is great enough so that the ψ_1 field lags the ψ_3 field by one-half wavelength and the 250-mb trough lies above the 750-mb ridge. In that case the thickness field $\psi_1 - \psi_3$ will lag the 500-mb geopotential field by 90° phase (one-quarter wavelength) so that the thermal advection will be a maximum. If the trough and ridge axes were tilted eastward with height, the vertical circulation would be reversed and the divergent motions would cause the vorticity extreme to decay in amplitude.

The remaining terms in (8.34) are more easily discussed if we neglect temperature advection effects by assuming that the ψ_1 and ψ_3 fields differ only in magnitude, not in phase, so that the motions are equivalent barotropic in nature. In this case ω_2 will still in general be nonzero due to the difference between the vorticity advections at 250 and 750 mb. This mechanism is expressed in the last two terms of (8.34). In an equivalent barotropic flow a nonzero ω_2 is necessary to maintain the phase relationship between the vorticity fields at 250 and 750 mb as those fields are advected along by the horizontal wind at each level.

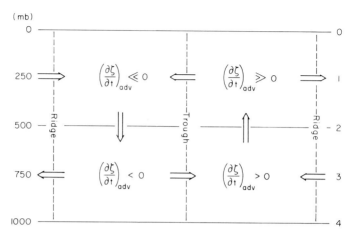

Fig. 8.6   The pattern of advective vorticity changes and the associated secondary circulation for equivalent barotropic flow.

In Fig. 8.6 the pattern of local vorticity change due to advection and the associated secondary circulation are shown schematically for the equivalent barotropic case. Since in general the nondivergent wind increases with height up to the tropopause the vorticity advection at 250 mb will be larger than that at 750 mb. As a consequence the vorticity tendencies at 250 mb will be larger so that

$$\frac{\partial}{\partial t}(\mathbf{V}^2\psi_1 - \mathbf{V}^2\psi_3) \begin{array}{l} > 0 \\ < 0 \end{array} \qquad \begin{array}{l} \text{east of the trough} \\ \text{west of the trough} \end{array}$$

which implies that

$$\frac{\partial}{\partial t}(\psi_1 - \psi_3) \begin{array}{l} < 0 \\ > 0 \end{array} \qquad \begin{array}{l} \text{east of the trough} \\ \text{west of the trough} \end{array}$$

Thus, the 250–750-mb thickness must change in time as the vorticity fields are advected along. Since horizontal temperature advection is zero in an equivalent barotropic flow, only adiabatic cooling or warming due to vertical motion can account for the thickness change. Thus ω_2 must be nonzero in order to produce the required adiabatic temperature changes.

East of the trough the thickness is decreasing in time so that the vertical motion is upward, while west of the trough thickness is increasing so that there must be subsidence warming. The divergent motions associated with this vertical motion field will themselves influence the vorticity tendencies through the divergence term in the vorticity equation. In Fig. 8.7 the vorticity tendency due to these divergent motions is indicated schematically for an equivalent

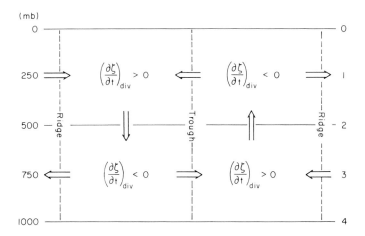

Fig. 8.7   The vorticity changes due to divergence associated with the secondary circulation in equivalent barotropic flow.

barotropic system. Comparing Figs. 8.6 and 8.7 we see that the divergence field acts to oppose the advective vorticity tendencies at the upper level, but at the lower level the divergent and advective tendencies are in phase. As a result the total vorticity tendency is less than the advective tendency at the upper level and greater than the advective tendency at the lower level:

$$\left| \frac{\partial}{\partial t} \nabla^2 \psi_1 \right| < \left| \frac{\partial}{\partial t} \nabla^2 \psi_1 \right|_{adv}$$

$$\left| \frac{\partial}{\partial t} \nabla^2 \psi_3 \right| > \left| \frac{\partial}{\partial t} \nabla^2 \psi_3 \right|_{adv}$$

The divergent motions thus *reduce* the differential vorticity tendencies, and thereby tend to slow down the eastward propagation of the vorticity wave at the upper level where advection is strong, and increase the speed of propagation of the vorticity wave at the lower level where advection is weak. Thus, the secondary circulation enables the system to remain equivalent barotropic as it is advected along at a speed intermediate between the zonal wind speed at the lower level and that at the upper level.

From a slightly different viewpoint, this secondary circulation may be regarded as a consequence of the continual adjustment of synoptic scale flow so that it remains in geostrophic and hydrostatic equilibrium. Thus in the previous example vorticity advection, by generating positive vorticity, tends to make the vorticity slightly supergeostrophic east of the trough at the upper level. The resulting unbalanced Coriolis force creates a divergent outflow from the region where the vorticity tendency is a maximum. By continuity, a vertical circulation is required to replace this diverging air. The adiabatic cooling associated with this vertical motion acts to deepen the geopotential trough at 250 mb so that the geopotential field tends to return to geostrophic balance with the wind. At the same time the divergence at 250 mb reduces the vorticity change. Thus, the departure from geostrophy is reduced in two ways by the secondary circulation: (1) by changing the thickness field through adiabatic cooling, and (2) by changing the vorticity through the divergence effect. In this manner the flow evolves always remaining in a state of quasi-geostrophic balance.

8.6   Numerical Solution of the Barotropic Vorticity Equation

So far in this chapter we have discussed three different filtered models, all of which are based on the quasi-geostrophic system, and all requiring the numerical solution of some type of vorticity equation in order to produce a forecast. In this section we briefly discuss the actual procedures involved in preparing a numerical prediction with the barotropic vorticity equation.

We wish to use the techniques of numerical analysis to obtain a finite-difference approximation to the vorticity equation. There are a number of problems associated with deriving finite-difference analogues to partial differential equations such as the vorticity equation. Some of the most serious problems relate to the *computational stability* of the finite-difference scheme. In order that a finite-difference scheme be computationally stable in the sense that a solution of the difference equations will approximate a solution of the original system, it turns out that the ratio of the time and space increments must satisfy certain conditions. In addition, truncation errors due to the approximate nature of derivatives estimated by finite differences are of concern. A detailed discussion of these problems is beyond the scope of this book. Only the rudiments of the necessary numerical analysis will be given here.

8.6.1   FINITE DIFFERENCING

The barotropic vorticity equation (8.16) can be rewritten in the following form:

$$\nabla^2 \chi + F(x, y, t) = 0 \qquad (8.35)$$

where

$$F(x, y, t) = \mathbf{V}_\psi \cdot \nabla(\nabla^2\psi + f) = \frac{\partial\psi}{\partial x}\frac{\partial}{\partial y}\nabla^2\psi - \frac{\partial\psi}{\partial y}\frac{\partial}{\partial x}\nabla^2\psi + \beta\frac{\partial\psi}{\partial x}$$

and

$$\chi = \frac{\partial \psi}{\partial t}$$

The advection of absolute vorticity, $F(x, y, t)$ may be calculated at any point in space provided that we know the field of $\psi(x, y, t)$. Equation (8.35) is a *Poisson equation* in the variable χ with $F(x, y, t)$ a known source function. It may be solved for χ by several standard methods. The most common method is to write (8.35) in finite-difference form and solve for χ approximately using the iterative method called *relaxation*.

Suppose that the horizontal x, y space is divided into a grid of $M \times N$ points separated by distance increments d. Then we can write the coordinate distances as $x = md$ and $y = nd$ where $m = 0, 1, 2, \ldots, M$ and $n = 0, 1, 2, \ldots, N$. Thus any point on the grid is uniquely identified by the indices (m, n). A portion of such a grid space is shown in Fig. 8.8.

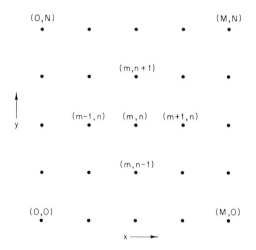

Fig. 8.8 Identification of mesh points on a finite difference grid of $M \times N$ points.

Using centered-difference formulas, derivatives of ψ at the point (m, n) may be expressed in terms of the values of ψ at surrounding points. For example,

$$\left(\frac{\partial \psi}{\partial x}\right)_{m,n} \simeq \frac{\psi_{m+1,n} - \psi_{m-1,n}}{2d}$$

$$\left(\frac{\partial \psi}{\partial y}\right)_{m,n} \simeq \frac{\psi_{m,n+1} - \psi_{m,n-1}}{2d}$$

In a similar fashion second derivatives at any point can be evaluated by taking the difference between the first derivatives evaluated at half grid increments on each side of the point:

$$\left(\frac{\partial^2 \psi}{\partial x^2}\right)_{m,n} \simeq \frac{(\partial\psi/\partial x)_{m+1/2,n} - (\partial\psi/\partial x)_{m-1/2,n}}{d}$$

$$\simeq \frac{\psi_{m+1,n} - \psi_{m,n}}{d^2} - \frac{\psi_{m,n} - \psi_{m-1,n}}{d^2}$$

$$= \frac{\psi_{m+1,n} - 2\psi_{m,n} + \psi_{m-1,n}}{d^2}$$

Similarly,

$$\left(\frac{\partial^2 \psi}{\partial y^2}\right)_{m,n} \simeq \frac{\psi_{m,n+1} - 2\psi_{m,n} + \psi_{m,n-1}}{d^2}$$

Thus, we may write as the finite-difference approximation to the horizontal Laplacian:

$$\nabla^2 \psi \simeq \frac{\psi_{m+1,n} + \psi_{m,n+1} + \psi_{m-1,n} + \psi_{m,n-1} - 4\psi_{m,n}}{d^2}$$

$$\equiv \overset{\blacktriangledown}{\nabla}^2 \psi_{m,n} \tag{8.36}$$

The finite-difference form of the Laplacian is simply proportional to the difference between the value of ψ at the central point and the average value at the four surrounding points.

We next write the term $F(x, y)$ in (8.35) in finite-difference form. However, before doing so we observe that if $F(x, y)$ is integrated over a closed region bounded by a line along which ψ is constant, it is easily verified that the average value of $F(x, y)$ over that region is zero. This would be the case, for example, if we integrated over the entire Northern Hemisphere and assumed that there was no flow across the equator. Hence, from (8.35) the average vorticity over the region is conserved.

Since $F(x, y)$ has an average value of zero, it is desirable that any finite-difference analogue to $F(x, y)$ also have zero average value. Otherwise, the average vorticity will not be conserved in the finite-difference form of the vorticity equation. In fact there are other integral constraints which motions governed by (8.35) must satisfy. In addition to conservation of average vorticity, kinetic energy and mean square vorticity are also conserved. Finite-difference schemes which conserve all these integral invariants have been designed. However, such treatments are beyond the scope of this discussion. It is sufficient here to indicate that by finite differencing $F(x, y)$ using the form

$$F(x, y) = \frac{\partial}{\partial y}\left(\frac{\partial \psi}{\partial x}\nabla^2\psi\right) - \frac{\partial}{\partial x}\left(\frac{\partial \psi}{\partial y}\nabla^2\psi\right) + \beta\frac{\partial \psi}{\partial x} \qquad (8.37)$$

it is possible to obtain a finite-difference analogue in which both the energy and average vorticity are conserved on the grid mesh. Thus, as a finite-difference form for $F(x, y)$ we take centered differences in (8.37) to get

$$F_{m,n} = \frac{1}{4d^2}\left[(\psi_{m+1,n+1} - \psi_{m-1,n+1})\nabla^2\psi_{m,n+1}\right.$$

$$- (\psi_{m+1,n-1} - \psi_{m-1,n-1})\nabla^2\psi_{m,n-1}$$

$$- (\psi_{m+1,n+1} - \psi_{m+1,n-1})\nabla^2\psi_{m+1,n}$$

$$\left. - (\psi_{m-1,n+1} - \psi_{m-1,n-1})\nabla^2\psi_{m-1,n}\right]$$

$$+ \frac{\beta}{2d}(\psi_{m+1,n} - \psi_{m-1,n}) \qquad (8.38)$$

It is readily verified that

$$\sum_{m-1}^{M-1}\sum_{n=1}^{N-1}F_{m,n} = 0$$

provided that ψ is constant on the boundaries. Therefore, the form of the advection term given in (8.38) conserves the average vorticity. In a similar fashion it is possible to demonstrate that (8.38) also conserves the average kinetic energy. Thus, the finite-difference form of (8.35) may be written as

$$\nabla^2\chi_{m,n} + F_{m,n} = 0 \qquad (8.39)$$

8.6.2   MAP FACTORS

Before discussing the relaxation method of inverting the Laplacian in (8.39) to obtain $\chi_{m,n}$ it is appropriate to digress slightly and discuss the question of map factors. It has been assumed above that the points on a finite-difference grid represent points in space separated by equal distances along the surface of the earth. In reality finite-difference grids are often square grids set on a map which is a flat projection of part of the curved earth. Usually meteorologists use *conformal* maps which preserve the shape in mapping any small area element of the earth, but may distort the relative size of the element. Thus, mesh points of a square grid on a conformal map will not in general correspond to uniform spacing on the real earth. An example of such a map is shown in Fig. 8.9 which shows a *polar stereographic* projection.

If the grid distance d_s on the map is measured in units which correspond to the actual grid distance at the latitude where the map projection is true,

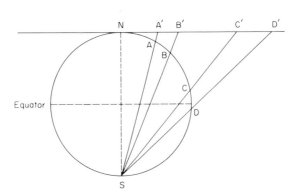

Fig. 8.9   Geometry of the polar stereographic projection. The arcs AB and CD on the surface of the earth map onto the lines $A'B'$ and $C'D'$, respectively.

then the *map factor* μ may be defined as the ratio of the map grid distance to the actual grid distance at any other latitude. For a polar stereographic projection, true at 90°N it can be shown that

$$\mu = \frac{d_s}{d} = \frac{2}{(1 + \sin \phi)}$$

In this case a given grid distance d_s will represent twice as much real distance at the pole as at the equator. Therefore, in order to preserve the accuracy of finite difference formulae, we must replace d by d_s/μ in (8.36) and (8.38) when the grid is referred to a map projection.

8.6.3   RELAXATION

We now consider the problem of solving (8.39) for $\chi_{m,n}$. Formally, we can write the solution as $\chi_{m,n} = -\nabla^{-2}F_{m,n}$ merely by inverting the finite-difference Laplacian operator. However, in order to actually invert this operator by any standard method, we must first specify suitable boundary conditions on χ along a closed curve. For our purposes, it is sufficient to assume that ψ is zero for all time at all boundary points.

The most practical scheme for solving (8.39) on a large grid mesh is an iterative technique known as *relaxation*. Relaxation schemes may be divided into two classes: (1) simultaneous relaxation, and (2) successive relaxation. These two methods will be discussed in turn below.

Since relaxation is an iterative procedure, some method is needed to identify the order of approximation. We thus label χ with the superscript ν to indicate the νth guess χ^ν. If the method is convergent, χ^ν should approach the true solution χ at all points as $\nu \to \infty$. Using this notation we write the initial guess for the χ field as $\chi^0_{m,n}$. Substituting the initial guess into (8.39) we obtain

$$\chi^0_{m+1,n} + \chi^0_{m-1,n} + \chi^0_{m,n+1} + \chi^0_{m,n-1} - 4\chi^0 = -d^2 F_{m,n} + R^0_{m,n} \qquad (8.40)$$

where $R^0_{m,n}$ is the *residual*, which is a measure of the difference between the initial guess and the true solution. The residual at any point (m,n) may be reduced to zero by altering the guess $\chi^0_{m,n}$ at that point to $\chi^1_{m,n}$ defined by

$$\chi^1_{m,n} = \chi^0_{m,n} + \tfrac{1}{4}R^0_{m,n} \qquad (8.41)$$

while leaving the guesses at all surrounding points unchanged. This can be seen by substituting back into (8.40). If the correction formula (8.41) is applied at all interior grid points, the result will be a new estimate of the field designated as $\chi^1_{m,n}$. The residuals can now be computed again by generalizing (8.40) to any order of approximation v: Thus,

$$R^v_{m,n} = d^2 F_{m,n} + \nabla^2 \chi^v_{m,n} \qquad (8.42)$$

The $(v + 1)$th guess can then be computed according to

$$\chi^{v+1}_{m,n} = \chi^v_{m,n} + \tfrac{1}{4}R^v_{m,n} \qquad (8.43)$$

This scheme is called simultaneous relaxation because the entire new field $\chi^{v+1}_{m,n}$ is guessed using residuals computed from the old field.

However, it is clear that once a new guess has been made at a given point, the new values can be used to modify the residuals at the surrounding points. Thus, the residuals can be computed sequentially starting from grid point 1,1 and working to the right along the grid to point $M - 1, 1$, then skipping to the second interior row of points and working from point 1, 2 to $M - 1, 2$, etc., as shown in Fig. 8.10. The formula for computing residuals then becomes

$$R^v_{m,n} = d^2 F_{m,n} + \chi^v_{m+1,n} + \chi^v_{m,n+1} + \chi^{v+1}_{m-1,n} + \chi^{v+1}_{m,n-1} - 4\chi^v_{m,n} \qquad (8.44)$$

Thus as shown in Fig. 8.10 at each point we use two new guesses and three old guesses to obtain the residual.

Although a general discussion of the conditions for convergence and rates of convergence for relaxation techniques is quite complicated, a simple example will suffice to indicate the superiority of successive relaxation to simultaneous relaxation. We consider the example shown in Fig. 8.11 in which we solve the finite-difference equation

$$\nabla^2 \chi + 1 = 0 \qquad (8.45)$$

for a mesh on which there are only two interior grid points labeled 1 and 2. The value of χ is taken to be zero on all boundary points. Applying formula

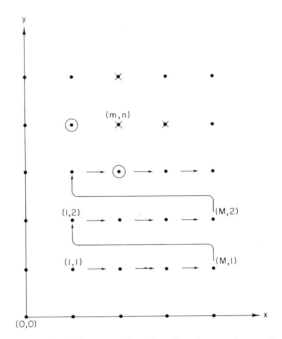

Fig. 8.10   Arrangement of grid for successive relaxation. Arrows show order of progression through the grid. The residual at (m, n) is computed using old guesses at the points indicated with $\times$ and new guesses at the circled points.

(8.45) we find that simultaneous relaxation gives for the $(\nu + 1)$th guess at the interior points

$$-4\chi_1^{\nu+1} + \chi_2^{\nu} = -1$$
$$-4\chi_2^{\nu+1} + \chi_1^{\nu} = -1$$

(8.46)

Now since χ^{ν} is the νth approximation to the true solution χ, we can express the error in the νth approximation as $\varepsilon^{\nu} = \chi^{\nu} - \chi$. Substituting this form into (8.46) and observing that χ satisfies (8.46) exactly we get

$$-4\varepsilon_1^{\nu+1} + \varepsilon_2^{\nu} = 0$$
$$-4\varepsilon_2^{\nu+1} + \varepsilon_1^{\nu} = 0$$

(8.47)

Since (8.47) is a recursive formula valid for all ν, we can write

$$-4\varepsilon_1^{\nu+2} + \varepsilon_2^{\nu+1} = 0$$

which, upon substituting for $\varepsilon_2^{\nu+1}$ from (8.47), gives

$$\varepsilon_1^{\nu+2} = \tfrac{1}{16}\varepsilon_1^{\nu}$$

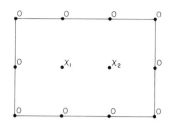

Fig. 8.11 A grid with two interior points.

Thus, the error in this simple example decreases to $\frac{1}{16}$ of its initial value in two interations of simultaneous relaxation.

In the successive method, on the other hand, the $(v + 1)$th guess is given by

$$-4\chi_1^{v+1} + \chi_2^{v} = 1$$
$$-4\chi_2^{v+1} + \chi_1^{v+1} = 1$$

In this case the errors satisfy the equations

$$-4\varepsilon_1^{v+1} + \varepsilon_2^{v} = 0$$
$$\varepsilon_1^{v+1} - 4\varepsilon_2^{v+1} = 0$$

from which we find

$$\varepsilon_1^{v+1} = \tfrac{1}{16}\varepsilon_1^{v}$$

Hence, the error decreases twice as fast as in the case of simultaneous relaxation. For more realistic cases involving a larger number of grid points, the convergence of the relaxation procedure would not be so rapid. However, in all cases it is found that using the successive method will improve the rate of convergence. Further improvement is possible by systematically *over-relaxing*. Over-relaxation merely involves multiplying the residual by a factor somewhat greater than $\frac{1}{4}$ in computing the new guess. In practice, only a few iterations are usually required for adequate convergence, provided that the initial guess is reasonably good.

8.6.4 TIME INTEGRATION

In the previous subsections we have shown how to write the space derivatives in the vorticity equation in finite difference form and how to use relaxation to obtain the $\chi_{m,n}$ field at any time given the $F_{m,n}$ field at that time. To forecast the future circulation we must now extrapolate ahead in time using a finite-difference approximation. Choosing a centered-differencing scheme we may write

$$\psi(t_0 + \delta t) = \psi(t_0 - \delta t) + 2\,\delta t\,\chi(t_0)$$

This scheme requires knowledge of ψ at two time levels, $\psi(t_0 - \delta t)$ and $\psi(t_0)$ in order to compute $\psi(t_0 + \delta t)$. Since at the first time step of the forecast $(t_0 = 0)$ only $\psi(t_0)$ is known, we must use a forward time step to initiate the forecast. Thus,

$$\psi(\delta t) = \psi(0) + \delta t\, \chi(0)$$

Specification of the time increment δt turns out to be a matter of crucial importance. Obviously, a larger value of δt will tend to give a poor approximation because in effect the tendency is linearly extrapolated over the entire interval $2\,\delta t$. Errors of this type which arise from the fact that finite differences are only approximations to the actual derivatives are called *truncation errors*. Such errors can be controlled by making δt sufficiently small. Of far greater significance, however, is the phenomenon of *computational instability*. It turns out that the space increment d and the time increment δt cannot be selected independently. They must, rather, be chosen so that the fastest traveling wave in the solution moves less than one grid interval in one time increment. For a two-dimensional grid, if the wave speed is designated by c, it can be shown that δt and d must satisfy

$$\frac{c\,\delta t}{d} < \frac{1}{\sqrt{2}} \tag{8.48}$$

If the criterion (8.48) is not satisfied, spurious wave modes are introduced which rapidly amplify so that the computed ψ field soon bears no resemblence to the true solution.

The existence of computational instability is one of the prime motivations for using filtered equations. In the quasi-geostrophic system no gravity or sound waves occur. Thus, the speed c in (8.48) is just the maximum wind speed. Typically, $c < 50$ m sec^{-1} so that for a grid interval of 200 km a time increment of over one hour would be permissible. On the other hand, if we use the complete equations, c must be set equal to the speed of sound which is the fastest wave described by the equations. Thus, $c \simeq 300$ m sec^{-1} and for a 200-km grid interval a time step of only a few minutes is permitted. Since the time scale for significant meteorological changes is much longer than a few minutes, it is generally more economical to use the filtered equations.

The procedure for preparing a numerical forecast with the barotropic vorticity equation can now be summarized as follows:

1. Use the observed geopotential field at time $t = 0$ to compute $\psi_{m,n}(0)$.
2. Evaluate $F_{m,n}$ at all grid points
3. Use relaxation to solve (8.35) for the tendency field $\chi_{m,n}$.
4. Extrapolate ahead a time increment δt to obtain $\psi_{m,n}(t + \delta t)$ using centered differencing except at the first step when a forward difference must be used.

5. Use the predicted field $\psi_{m,n}(t + \delta t)$ as data and repeat steps 2 through 4 until the desired forecast period is reached. Thus for a 24-hr forecast 24 time steps would be required with a one-hour time increment.

6. Use the forecast $\psi_{m,n}$ field to compute the predicted geopotential and nondivergent wind fields.

8.7 Primitive Equation Models

The filtered models discussed in the previous sections of this chapter have all involved the various approximations of the quasi-geostrophic system. Although these approximations are generally valid to within about 10–20 percent in middle latitudes, they do place a definite limit on the accuracy of the forecast. Furthermore, if these models are to be used in low latitudes, the streamfunction cannot be determined from the geostrophic vorticity field but must be obtained from the more complicated balance equation (8.74). This is a nonlinear equation which is difficult and time consuming to solve at each time step. Therefore, it is worthwhile to investigate the possibility of using less restrictive modeling assumptions.

Our scale analysis (Section 2.4) indicated that the hydrostatic approxima-tion is far more accurate than the other filtering approximations. Thus, we will continue to assume that the motions are hydrostatic. The equations can then be written in pressure coordinates. The resulting system then consists of three prognostic equations (the x and y components of the momentum equation, and the thermodynamic energy equation) and three diagnostic equations (the continuity equation, the hydrostatic approximation, and the equation of state). These constitute a closed set in the dependent variables u, v, ω, ϕ, α, and θ which is referred to as the *primitive equations*.

8.7.1 SIGMA COORDINATES

The isobaric coordinate system has many advantages which have been previously mentioned. Among these are the following: (1) Meteorological data is normally referred to isobaric surfaces. (2) The continuity equation has a simple form. (3) Density does not explicitly appear. (4) Sound waves are completely filtered. However, these advantages are partially offset by the fact that the boundary condition at the ground is difficult to handle in the isobaric system. In fact we have generally used the approximate condition

$$\omega(p_0) \simeq -\rho_0 g w(z_0)$$

as the lower boundary condition. Here we have assumed that the height of the ground z_0 is coincident with the pressure surface p_0 (where p_0 is usually set equal to 1000 mb). These assumptions are of course not strictly valid even when the ground is level. Pressure does change at the ground. But more importantly, the height of the ground generally varies so that even if the pressure tendency were zero everywhere, the lower boundary condition should not be applied at a constant p_0. Rather, we should set $p_0 = p_0(x, y)$. However, it is very inconvenient for mathematical analysis to have a boundary condition which must be applied at a surface which is a function of the horizontal variables. Therefore, a modification of the pressure coordinate system, called the *sigma* system is usually used in numerical modeling. In the sigma system the vertical coordinate is the pressure normalized with the surface pressure.[2]

$$\sigma \equiv \frac{p}{p_s} \qquad (8.49)$$

where $p_s(x, y, t)$ is the pressure at the surface. Thus, σ is a *nondimensional* independent vertical coordinate which decreases upward from a value $\sigma = 1$ at the ground to $\sigma = 0$ at the top of the atmosphere. Hence, in sigma coordinates the lower boundary condition will always apply exactly at $\sigma = 1$. Furthermore, the vertical σ velocity

$$\dot{\sigma} \equiv \frac{d\sigma}{dt}$$

will always be zero at the ground even in the presence of sloping terrain. Thus, the lower boundary condition in the σ system is merely

$$\dot{\sigma} = 0 \qquad \text{at} \quad \sigma = 1 \qquad (8.50)$$

It is now necessary to transform the primitive equations from the (x, y, p) system to the (x, y, σ) system. In the (x, y, p) system, the basic primitive equations can be written in vector notation as

$$\frac{d\mathbf{V}}{dt} + f\mathbf{k} \times \mathbf{V} = -\nabla\Phi \qquad (8.51)$$

$$\nabla \cdot \mathbf{V} + \frac{\partial \omega}{\partial p} = 0 \qquad (8.52)$$

$$\frac{\partial \Phi}{\partial p} + \alpha = 0 \qquad (8.53)$$

$$c_p \frac{d \ln \theta}{dt} = \frac{dQ}{dt} \qquad (8.54)$$

$$\theta = \left(\frac{p_0}{p}\right)^{R/c_p} \left(\frac{p\alpha}{R}\right) \qquad (8.55)$$

[2] It is important not to confuse the independent variable σ in this section with the static stability parameter of the quasi-geostrophic system which has also been labeled σ.

where

$$\frac{d}{dt} = \frac{\partial}{\partial t} + \mathbf{V} \cdot \mathbf{\nabla} + \omega \frac{\partial}{\partial p}$$

and the horizontal $\mathbf{V}$ operator refers to differentiation at constant pressure. Since in general the σ and p surfaces will not coincide, partial derivatives evaluated at constant σ will not be equal to the analogous derivatives evaluated at constant p. The transformation from p to σ coordinates can be derived in a manner similar to the transformation from z to p coordinates discussed in Section 3.1.

We consider the transformation of the x component of the gradient of geopotential. From Fig. 8.12 we see that

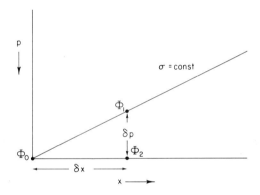

Fig. 8.12 The transformation from p coordinates to σ coordinates.

$$\frac{\Phi_1 - \Phi_0}{\delta x} = \frac{\Phi_2 - \Phi_0}{\delta x} - \frac{\Phi_2 - \Phi_1}{\delta x}$$

$$= \frac{\Phi_2 - \Phi_0}{\delta x} - \frac{\Phi_2 - \Phi_1}{\delta \sigma} \frac{\delta \sigma}{\delta x} \tag{8.56}$$

where $\delta \sigma = (\partial \sigma / \partial x)_p \, \delta x$. Thus, taking the limits δx, $\delta \sigma \to 0$ (8.56) yields

$$\left(\frac{\partial \Phi}{\partial x}\right)_\sigma = \left(\frac{\partial \Phi}{\partial x}\right)_p - \frac{\partial \Phi}{\partial \sigma} \left(\frac{\partial \sigma}{\partial x}\right)_p$$

which with the aid of (8.49) may be written

$$\left(\frac{\partial \Phi}{\partial x}\right)_\sigma = \left(\frac{\partial \Phi}{\partial x}\right)_p + \frac{\sigma}{p_s} \frac{\partial p_s}{\partial x} \frac{\partial \Phi}{\partial \sigma}$$

Since any other variable will transform in an analogous way we can write the general transformation as

$$\mathbf{V}_p(\ ) = \mathbf{V}_\sigma(\ ) - \frac{\sigma}{p_s} \mathbf{V}p_s \frac{\partial(\ )}{\partial \sigma} \qquad (8.57)$$

Applying the transformation formula (8.57) to the momentum equation (8.51) we get

$$\frac{d\mathbf{V}}{dt} + f\mathbf{k} \times \mathbf{V} = -\mathbf{V}\Phi + \frac{\sigma}{p_s} \mathbf{V}p_s \frac{\partial \Phi}{\partial \sigma} \qquad (8.58)$$

where $\mathbf{V}$ is now applied holding σ *constant*, $\mathbf{V}$ is the horizontal velocity measured along σ surfaces, and the total differential is

$$\frac{d}{dt} = \frac{\partial}{\partial t} + \mathbf{V} \cdot \mathbf{V} + \dot\sigma \frac{\partial}{\partial \sigma}$$

The equation of continuity can be transformed to the σ system as follows: From (8.57) we write the horizontal divergence as

$$(\mathbf{V} \cdot \mathbf{V})_p = (\mathbf{V} \cdot \mathbf{V})_\sigma - \frac{\sigma}{p_s} \frac{\partial \mathbf{V}}{\partial \sigma} \cdot \mathbf{V}p_s \qquad (8.59)$$

To transform the term $\partial\omega/\partial p$ we first note that from (8.49) since p_s does not depend on σ

$$\frac{\partial}{\partial p} = \frac{\partial}{\partial(\sigma p_s)} = \frac{1}{p_s} \frac{\partial}{\partial \sigma}$$

Thus, the continuity equation can be written

$$p_s(\mathbf{V} \cdot \mathbf{V})_p + \frac{\partial\omega}{\partial \sigma} = 0 \qquad (8.60)$$

Now the sigma vertical velocity $\dot\sigma$ can be written as

$$\dot\sigma \equiv \frac{d\sigma}{dt} = \left(\frac{\partial\sigma}{\partial t} + \mathbf{V} \cdot \mathbf{V}\sigma\right)_p + \omega \frac{\partial\sigma}{\partial p}$$

$$= -\frac{\sigma}{p_s}\left(\frac{\partial p_s}{\partial t} + \mathbf{V} \cdot \mathbf{V}p_s\right) + \frac{\omega}{p_s}$$

Differentiating the above with respect to σ and rearranging terms yields

$$\frac{\partial\omega}{\partial \sigma} = p_s \frac{\partial\dot\sigma}{\partial \sigma} + \left(\frac{\partial p_s}{\partial t} + \mathbf{V} \cdot \mathbf{V}p_s\right)_\sigma + \sigma \frac{\partial \mathbf{V}}{\partial \sigma} \cdot \mathbf{V}p_s \qquad (8.61)$$

Substituting from (8.59) and (8.61) into (8.60) we obtain the transformed continuity equation

$$\mathbf{V} \cdot (p_s \mathbf{V}) + p_s \frac{\partial \dot{\sigma}}{\partial \sigma} + \frac{\partial p_s}{\partial t} = 0 \qquad (8.62)$$

With the aid of the equation of state the hydrostatic approximation (8.53) may be written in the sigma system as

$$\frac{\partial \Phi}{\partial \sigma} = -\frac{RT}{\sigma} \qquad (8.63)$$

where the temperature T is related to θ by Poisson's equation

$$\theta = T\left(\frac{p_0}{\sigma p_s}\right)^{R/c_p} \qquad (8.64)$$

with $p_0 = 1000$ mb.

The set of equations (8.54), (8.58), (8.62), (8.63), and (8.64) contains the six dependent variables $\mathbf{V}$, $\dot{\sigma}$, θ, T, Φ, p_s. We, therefore, need an additional equation to make the system complete. This relation is the surface pressure tendency equation which can be readily obtained by integrating the continuity equation (8.62) vertically and using the boundary conditions that

$$\dot{\sigma} = 0 \qquad \text{at} \quad \sigma = 0, 1$$

The result is

$$\frac{\partial p_s}{\partial t} = -\int_0^1 \mathbf{V} \cdot (p_s \mathbf{V}) \, d\sigma \qquad (8.65)$$

Equation (8.65) simply states that the rate of increase of the surface pressure at a given point equals the mass convergence in a unit cross-section column above the point. With the inclusion of (8.65) we now have a complete set of prediction equations which can be written in finite-difference form and numerically integrated.

8.7.2   A Two-Layer Primitive Equation Model

One important way in which primitive equation models differ from models based on the quasi-geostrophic system is that the static stability is generally a specified parameter in the latter type, whereas in the former the static stability must be allowed to vary in space and time. Thus, in primitive equation models it is necessary to compute $\partial\theta/\partial p$ explicitly at each time step.[3] Temperature

---

[3] It can be shown that if $\partial\theta/\partial p$ is specified as a function of pressure alone it is not possible to formulate a primitive equation model which satisfies the requirements of energy conservation.

must then be predicted at a minimum of *two* data levels rather than carried at a single level as in the two parameter quasi-geostrophic model described in Section 8.5. A simple vertical differencing scheme which satisfies this requirement is shown in Fig. 8.13.

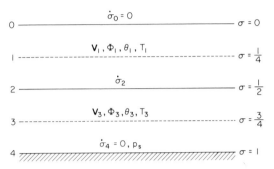

Fig. 8.13   Vertical differencing scheme for a two-level primitive equation model.

As in the two-parameter model we divide the atmosphere into four layers separated by surfaces labeled 0–4. The momentum equation and thermodynamic energy equation are applied at levels 1 and 3. The vertical differencing of these equations requires knowledge of the sigma vertical motion at level 2. The field of $\dot{\sigma}_2$ may be obtained diagnostically through use of the continuity equation as follows: Writing the continuity equation for levels 1 and 3 with vertical derivatives replaced by centered differences, we obtain

$$\frac{\partial p_s}{\partial t} + \mathbf{V} \cdot (p_s \mathbf{V}_1) + 2p_s \dot{\sigma}_2 = 0 \qquad (8.66)$$

$$\frac{\partial p_s}{\partial t} + \mathbf{V} \cdot (p_s \mathbf{V}_3) - 2p_s \dot{\sigma}_2 = 0 \qquad (8.67)$$

Adding these two equations gives a finite-difference form of the tendency equation (8.65),

$$\frac{\partial p_s}{\partial t} + \tfrac{1}{2}\mathbf{V} \cdot [p_s(\mathbf{V}_1 + \mathbf{V}_3)] = 0 \qquad (8.68)$$

while subtracting (8.67) from (8.66) yields the desired diagnostic equation for $\dot{\sigma}_2$,

$$\dot{\sigma}_2 = \frac{-1}{4p_s} \mathbf{V} \cdot [p_s(\mathbf{V}_1 - \mathbf{V}_3)] \qquad (8.69)$$

In a similar manner, the geopotentials Φ_1 and Φ_3 may be obtained diagnostically by vertical integration of the hydrostatic approximation (8.63) since T_1

and T_3 are known functions of θ_1 and θ_3, and Φ_s, the geopotential at the surface, is a known function of x and y.

To summarize, the procedure for forecasting with a two-level primitive equation model is as follows:

1. Write suitable finite difference analogues of the momentum and thermodynamic energy equations at levels 1 and 3, and the surface pressure tendency equation.

2. Use these prognostic equations to obtain the tendencies of the V_1, V_3, θ_1, θ_3, and p_s fields.

3. Extrapolate the tendencies ahead using a suitable time differencing scheme.

4. Use the new values of the V_1, V_3, θ_1, θ_3, and p_s fields to diagnostically determine $\dot{\sigma}_2$, Φ_1, and Φ_3.

5. Repeat steps 2–4 until the desired forecast time is reached.

In applying this scheme, it must be noted that the primitive equations in sigma coordinates contain the mechanisms for both gravity-wave and horizontal acoustic-wave propagation. It is thus necessary that the time increment be kept small enough to satisfy the condition

$$\frac{c\,\Delta t}{d} < \frac{1}{\sqrt{2}}$$

where c is the speed of a horizontally propagating acoustic wave which is the fastest-moving wave solution for these equations. For this reason, time steps in primitive equation forecasts must be considerably shorter than those allowed for a quasi-geostrophic model with equal horizontal resolution.

8.7.3 INITIAL DATA FOR THE PRIMITIVE EQUATIONS

In discussing Richardson's early attempt at numerical forecasting, it was mentioned that one reason for his poor results was the lack of suitable initial data. Although data coverage has vastly improved since Richardson's time, measured wind velocities still are not accurate enough to allow us to use observed winds and pressures as initial data without some adjustments.

The basic problem in specifying initial data for primitive equation models can be illustrated by considering the relative magnitudes of the various terms in the momentum equation in pressure coordinates

$$\frac{\partial V}{\partial t} + (V \cdot \nabla)V + \omega \frac{\partial V}{\partial p} = -f k \times V - \nabla \Phi \qquad (8.70)$$

For synoptic scale motions the wind and pressure fields are in approximate geostrophic balance. Thus, the acceleration following the motion is measured by the small difference between the two nearly equal terms $f\mathbf{k} \times \mathbf{V}$ and $-\nabla\Phi$. Although the Φ field can be determined observationally with quite good accuracy, observed winds are often 10–20 percent in error. This means that by using observed winds in the initial data the estimated Coriolis force may be 10–20 percent in error at the initial time. Since the acceleration is normally only about 10 percent of the Coriolis force in magnitude, an acceleration computed using the observed wind and geopotential fields will generally be 100 percent in error. Such spurious accelerations not only lead to poor estimates of the initial pressure and velocity tendencies but also may produce large-amplitude gravity-wave oscillations as the flow attempts to adjust from the initial unbalanced state back toward a state of geostrophic balance. These strong gravity waves are not present in nature, but their presence in the solution of the model equations will quickly spoil any chance of a reasonable forecast.

One possible approach to avoid this problem might be to neglect the observed wind data, and derive a wind field from the observed Φ field. The simplest scheme of this type would be to assume that the initial wind field was in geostrophic balance. However, it can be shown that the error in the computed, initial *local* acceleration would still be ~ 100 percent. This can be seen by considering the example of stationary flow about a circular pressure system. In that case $\partial\mathbf{V}/\partial t = 0$ so that if we neglect the effects of vertical advection the flow will be in gradient wind balance (see Section 3.2.4). In vectorial form this balance is simply

$$(\mathbf{V} \cdot \nabla)\mathbf{V} + f\mathbf{k} \times \mathbf{V} = -\nabla\Phi \qquad (8.71)$$

However, if $\mathbf{V}$ were replaced by $\mathbf{V}_g$ in (8.70) the Coriolis and pressure gradient forces would identically balance. Thus, the inertial force would be unbalanced and

$$\left(\frac{\partial\mathbf{V}}{\partial t}\right)_{t=0} \simeq -(\mathbf{V}_g \cdot \nabla)\mathbf{V}_g$$

Therefore, by assuming that initially the wind is in geostrophic balance rather than gradient balance, we compute a local acceleration which is completely erroneous.

It should now be clear that in order to avoid large errors in the initial acceleration, the initial wind field should be determined by the gradient wind balance, not by the geostrophic approximation or direct observations. The gradient wind formula (3.15) is not in itself a suitable balance condition to use, because the radius of curvature must be computed for parcel trajectories. However, an equivalent balance condition can be obtained from the divergence equation (8.6). If we assume that *initially*

$$\frac{\partial}{\partial t}(\nabla \cdot \mathbf{V}) = 0 \qquad \text{and} \quad \nabla \cdot \mathbf{V} = 0$$

(8.6) can be written for time $t = 0$ as

$$\nabla^2 \left(\Phi + \frac{\mathbf{V} \cdot \mathbf{V}}{2} \right) = \nabla \cdot [\mathbf{k} \times \mathbf{V}(\zeta + f)] \qquad (8.72)$$

Equation (8.72) is, of course, just the horizontal divergence of (8.71). Since we have assumed that the initial velocity is nondivergent, we can replace $\mathbf{V}$ by the streamfunction ψ by letting

$$\mathbf{V}_\psi = \mathbf{k} \times \nabla \psi \qquad (8.73)$$

The resulting relationship between ψ and Φ is called the *balance equation*:

$$\nabla^2 [\Phi + \tfrac{1}{2}(\nabla \psi)^2] = \nabla \cdot [(f + \nabla^2 \psi) \nabla \psi] \qquad (8.74)$$

Equation (8.74) is linear in Φ, but nonlinear in ψ. Thus if ψ is known (8.74) can be solved by relaxation to determine Φ. In the usual midlatitude case, however, Φ is known and more sophisticated iterative techniques must be used to solve for ψ. In fact, since (8.74) is quadratic in ψ, there will generally exist two solutions for the ψ field corresponding to a given Φ field. One solution will correspond to the normal gradient wind balance, the other to the anomalous balance. Once ψ is determined from (8.74), the initial velocity field is computed using (8.73). The velocity field determined in this manner will be in gradient balance with the geopotential field. The initial local accelerations are thus very small. But more importantly, the initial local rate of change of divergence is zero which assures that large amplitude gravity oscillations are not excited by the initial conditions.

8.7.4. THE NMC SIX-LAYER MODEL

For operational numerical forecasting the United States National Meteorological Center (NMC) uses a model with six prediction levels in the vertical. This model employs a modification of the sigma coordinate system in which four separate σ domains are defined. The vertical structure of the model is shown in Fig. 8.14. The coordinates σ_B, σ_T, and σ_S refer to the planetary boundary layer, the middle troposphere, and the stratosphere, respectively. The choice of separate coordinates for each of these regions allows better vertical resolution both in the boundary layer and in the stratosphere. In this model the tropopause is considered to be a material surface. Thus, an additional dependent variable p_2 the pressure at the tropopause must be predicted. However the basic scheme is similar to that described for the two-level primitive equation model. The prognostic equations are solved at levels $k = 0, 1, \ldots, 5$. The constant-θ layer at the top is included for computational reasons, but has no meteorological significance.

Initial data in this model must be derived by interpolating geopotential values from constant-pressure maps. Since this interpolation involves some

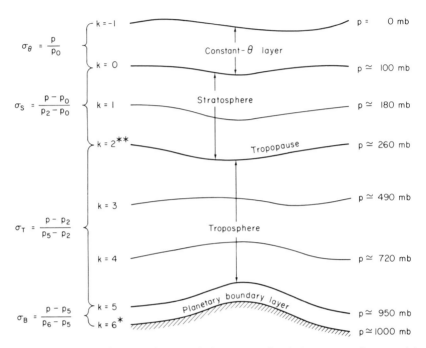

Fig. 8.14   Schematic diagram of the vertical structure of a six-layer σ coordinate model. (After Shuman and Hovermale, 1968.)

errors, it is necessary to determine the initial velocity field using the balance equation referred to sigma coordinates. The initial mass and velocity fields will then be in gradient balance on the sigma surfaces. For a grid distance of ~ 400 km, this model requires a time step of about ten minutes or less to avoid computational instability.

Problems

1. Compute the equivalent barotropic level for the following wind data:

p (mb)	u (m sec^{-1})	p (mb)	u (m sec^{-1})
1000	1.6	400	13.2
900	3.7	300	14.4
800	5.8	200	15.8
700	7.9	100	13.1
600	9.8	0	0
500	11.9		

2. Show that the nonlinear terms in the balance equation

$$G(x, y) \equiv -\nabla^2\left(\frac{\nabla\psi \cdot \nabla\psi}{2}\right) + \nabla \cdot (\nabla\psi \, \nabla^2\psi)$$

may be written in cartesian coordinates as

$$G(x, y) = 2\left[\frac{\partial^2\psi}{\partial x^2}\frac{\partial^2\psi}{\partial y^2} - \left(\frac{\partial^2\psi}{\partial x \, \partial y}\right)^2\right]$$

3. Express $G(x, y)$ from the previous problem in finite-difference form using centered differences.

4. Use relaxation to solve the Poisson equation

$$\nabla^2\chi_{m,n} + F_{m,n} = 0$$

for $\chi_{m,n}$ on the mesh points of the grid shown in Fig. 8.8 assuming $M = N = 5$, and that $\chi_{m,n}$ is zero on all boundary points. For an initial guess let $\chi^0_{m,n} = F_{m,n}$ where the values of $F_{m,n}$ at the interior points are as follows:

(m,n)	$F_{m,n}$	(m,n)	$F_{m,n}$
1, 1	1	3, 2	2
2, 1	4	1, 3	1
3, 1	1	2, 3	4
1, 2	2	3, 3	1
2, 2	6		

5. Show by using scale analysis that in the first approximation (8.5) and (8.6) reduce to (8.9) and (8.10) for midlatitude synoptic scale motions.

6. Show that the advection of absolute vorticity by the nondivergent wind

$$\mathbf{V}_\psi \cdot \nabla(\zeta + f)$$

can be written in the form given in (8.37).

7. Show by a formal transformation of variables from the (x, y, p) to the (x, y, σ) system that

$$\left(\frac{\partial}{\partial t} + \mathbf{V} \cdot \nabla\right)_p + \omega \frac{\partial}{\partial p} = \left(\frac{\partial}{\partial t} + \mathbf{V} \cdot \nabla\right)_\sigma + \dot{\sigma}\frac{\partial}{\partial \sigma}$$

Suggested References

Haltiner, *Numerical Weather Prediction*, presents an excellent treatment of the entire subject at graduate level.

Platzman (1967) presents an interesting historical review of Richardson's work and its relation to modern day forecasting methods.

Thompson, *Numerical Weather Analysis and Prediction*, covers much of the material in this chapter in greater depth. He presents a clear discussion of the various problems involved in using finite-difference formulas in numerical forecasting such as computational instability and truncation errors. Thompson's book assumes no previous acquaintance with finite-difference techniques.

Richtmeyer, *Difference Methods for Initial Value Problems*, is an advanced text which thoroughly treats the questions of truncation error and computational stability for a wide variety of problems.

Shuman and Hovermale (1968) present a complete description of the NMC operational six-layer primitive equation forecast model.

Phillips (1970), reviews several aspects of numerical modeling on an advanced level. Included is an interesting discussion of the reasons for the failure of Richardson's forecast. This article also includes an extensive list of references.

9 | Atmospheric Oscillations: Linear Perturbation Theory

In the previous chapter we discussed numerical techniques for solving the equations governing large-scale atmospheric motions. If the objective is to produce an accurate forecast of the circulation at some future time, a detailed numerical model based on the primitive equations and including processes such as latent heating, radiative transfer, and frictional dissipation should produce the best results. However, the inherent complexity of such a model generally precludes any simple interpretation of the physical processes which produce the predicted circulation. If we wish to gain physical insight into the fundamental nature of atmospheric motions, it is necessary to employ simplified models in which certain processes are omitted, and compare the results with those of more complete models. This is, of course, just what has been done in deriving the filtered quasi-geostrophic model. However, the quasi-geostrophic system still requires numerical solution of a complicated nonlinear system of equations. It is difficult to gain an appreciation for the processes essential to the development of baroclinic wave disturbances by examining the results of such numerical integrations.

In this chapter we will discuss a simple technique, the *perturbation method*, which is ideally suited for *qualitative* analysis of the nature of atmospheric motions. We then use perturbation theory to examine several types of pure waves in the atmosphere. In Chapter 10 perturbation theory will be applied to the problem of the development of synoptic wave disturbances.

9.1 The Perturbation Method

In the perturbation method all field variables are divided into two parts, a *basic state* portion which is usually assumed to be independent of time and longitude and a *perturbation* portion which is the local deviation of the field from the basic state. Thus, for example, if $\bar{u}$ designates a time and longitude averaged zonal velocity and u' is the deviation from that average, then the complete zonal velocity field is $u(x, t) = \bar{u} + u'(x, t)$. In that case, for example, the inertial acceleration $u \, \partial u / \partial x$ can be written

$$u \frac{\partial u}{\partial x} = (\bar{u} + u') \frac{\partial}{\partial x} (\bar{u} + u') = \bar{u} \frac{\partial u'}{\partial x} + u' \frac{\partial u'}{\partial x}$$

The basic assumptions of perturbation theory are that the basic states variables must themselves satisfy the governing equations when the perturbations are set to zero, and the perturbation fields must be small enough so that all terms in the governing equations which involve products of the perturbation variables can be neglected. The latter requirement would be met in the above example if $|u'/\bar{u}| \ll 1$ so that

$$\left| \bar{u} \frac{\partial u'}{\partial x} \right| \gg \left| u' \frac{\partial u'}{\partial x} \right|$$

By neglecting all terms which are nonlinear in the perturbations, the nonlinear governing equations are reduced to linear differential equations in the perturbation variables in which the basic state variables are specified coefficients. These equations can then be solved by standard methods to determine the character and structure of the perturbations in terms of the known basic state. Usually the perturbations are assumed to be sinusoidal waves and solution of the perturbation equations determines such characteristics as the propagation speed, vertical structure, and conditions for growth or decay of the waves. The perturbation technique is especially useful in studying the stability of a given basic state flow with respect to small superposed perturbations. This application will be the subject of Chapter 10.

9.2 Properties of Waves

The reason that it is useful to assume that the perturbations have the form of sinusoidal waves is that the linearized perturbation equations for atmospheric flow can often be combined to give a single equation which is a generalization of the so-called *wave equation*

$$\frac{\partial^2 \psi}{\partial t^2} = c^2 \frac{\partial^2 \psi}{\partial x^2} \tag{9.1}$$

An equation of the form (9.1) can be shown to have solutions corresponding to waves of arbitrary profile moving at the speed c in both the positive and negative x directions. We consider an arbitrary initial profile of the field ψ:

$$\psi = f(x) \quad \text{at} \quad t = 0$$

If this profile is translated in the positive x direction at speed c without change of shape then

$$\psi = f(x')$$

where x' is a coordinate moving at speed c so that $x = x' + ct$. Thus, in terms of the fixed coordinate x we can write

$$\psi = f(x - ct)$$

corresponding to a profile which moves in the positive x direction at speed c without change of shape. It may be easily verified (see Problem 1) that this function is indeed a solution of (9.1) as is the corresponding wave which moves in the negative x direction,

$$\psi = f(x + ct)$$

An example is shown in Fig. 9.1 for the case of a sinusoidal wave traveling in the positive direction, that is, $\psi = \sin(x - ct)$.

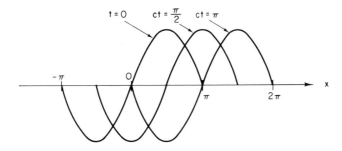

Fig. 9.1 A sinusoidal wave traveling in the positive x direction at speed c.

The representation of a perturbation as a simple sinusoidal wave might seem an oversimplification since disturbances in the atmosphere are never purely sinusoidal. It turns out, however, that any reasonably well-behaved function of longitude can be represented in terms of a zonal mean plus a *Fourier* series of sinusoidal components:

$$f(x) = \sum_{m=1}^{\infty} (A_m \sin k_m x + B_m \cos k_m x) \tag{9.2}$$

where $k_m = 2\pi m/L$ is the zonal wave number, L is the distance around a latitude circle, and m is an integer designating the number of waves around a circle of latitude. The coefficients A_m are calculated by multiplying both sides of (9.2) by $\sin(2\pi nx/L)$ and integrating around a latitude circle. Applying the orthogonality relationships:

$$\int_0^L \sin\frac{2\pi mx}{L}\sin\frac{2\pi nx}{L}\,dx = \begin{cases}0, & m \neq n\\L/2, & m = n\end{cases}$$

we obtain

$$A_m = \frac{2}{L}\int_0^L f(x)\sin\frac{2\pi mx}{L}\,dx$$

In a similar fashion, multiplying both sides in (9.2) by $\cos(2\pi nx/L)$ and integrating gives

$$B_m = \frac{2}{L}\int_0^L f(x)\cos\frac{2\pi mx}{L}\,dx$$

A_m and B_m are called the *Fourier coefficients*, whereas

$$f_m(x) = A_m \sin k_m x + B_m \cos k_m x \qquad (9.3)$$

is called the *m*th *Fourier* component or *mth harmonic* of the function $f(x)$. If the Fourier coefficients are computed for a quantity such as the longitudinal dependence of the observed geopotential perturbation, it turns out that the largest amplitude Fourier components will be those for which m is close to the observed number of troughs or ridges around a latitude circle. When only qualitative information is desired, it is usually sufficient to limit the analysis to a single typical Fourier component, and assume that the behavior of the actual field will be similar to that of the component. The expression for a Fourier component may be written more compactly by using complex exponential notation. According to the Euler formula

$$e^{ix} = \cos x + i \sin x$$

where $i \equiv \sqrt{-1}$ is the imaginary unit. Thus, we can write

$$f_m(x) = \mathrm{Re}\{C_m e^{ik_m x}\}$$
$$= \mathrm{Re}\{C_m \cos k_m x + iC_m \sin k_m x\} \qquad (9.4)$$

where $\mathrm{Re}\{\ \}$ denotes "real part of," and C_m is a complex coefficient. Comparing (9.3) and (9.4) we see that the two representations of $f_m(x)$ are identical provided that

$$B_m = \mathrm{Re}\{C_m\} \qquad \text{and} \qquad A_m = -\mathrm{Im}\{C_m\}$$

where Im{ } stands for "imaginary part of." This exponential notation will generally be used for applications of the perturbation theory below as well as in Chapter 10.

9.3 Sound Waves

To introduce the perturbation method we apply the technique first to certain pure types of wave disturbances which can propagate in the atmosphere. Sound waves are perhaps the simplest example to isolate. It was previously pointed out in Section 8.2 that sound waves are longitudinal waves which propagate as a result of alternating adiabatic compression and rarefaction due to the divergence field in the fluid. To exclude the possibility of *transverse* oscillations (that is, oscillations in which the particle motion is at right angles to the direction of phase propagation), we assume at the outset that the wave propagates in the x direction and that $v = w = 0$. In addition we eliminate all dependence on y and z by assuming that $u = u(x, t)$. With these restrictions the momentum equation, continuity equation, and thermodynamic energy equation for adiabatic motion are, respectively,

$$\frac{du}{dt} + \frac{1}{\rho}\frac{\partial p}{\partial x} = 0 \qquad (9.5)$$

$$\frac{d\rho}{dt} + \rho\frac{\partial u}{\partial x} = 0 \qquad (9.6)$$

$$\frac{d\ln\theta}{dt} = 0 \qquad (9.7)$$

where for this case

$$\frac{d}{dt} = \frac{\partial}{\partial t} + u\frac{\partial}{\partial x}$$

Recalling that

$$\theta = \frac{p}{\rho R}\left(\frac{p_0}{p}\right)^{R/c_p} \qquad \text{where} \quad p_0 = 1000 \quad \text{mb}$$

we may eliminate θ in (9.7) to give

$$\frac{1}{\gamma}\frac{d\ln p}{dt} - \frac{d\ln\rho}{dt} = 0 \qquad (9.8)$$

where $\gamma = c_p/c_v$. Eliminating ρ between (9.6) and (9.8) gives

$$\frac{1}{\gamma}\frac{d\ln p}{dt} + \frac{\partial u}{\partial x} = 0 \qquad (9.9)$$

The dependent variables are now divided into constant basic state portions (denoted by the overbars) and perturbation portions (denoted by primes):

$$u(x, t) = \bar{u} + u'(x, t)$$
$$p(x, t) = \bar{p} + p'(x, t) \qquad (9.10)$$
$$\rho(x, t) = \bar{\rho} + \rho'(x, t)$$

Substituting (9.10) into (9.5) and (9.9) we obtain

$$\frac{\partial}{\partial t}(\bar{u} + u') = (\bar{u} + u')\frac{\partial}{\partial x}(\bar{u} + u') + \frac{1}{\bar{\rho} + \rho'}\frac{\partial}{\partial x}(\bar{p} + p') = 0$$

$$\frac{\partial}{\partial t}(\bar{p} + p') + (\bar{u} + u')\frac{\partial}{\partial x}(\bar{p} + p') + \gamma(\bar{p} + p')\frac{\partial}{\partial x}(\bar{u} + u') = 0$$

We next observe that provided $|\rho'/\bar{\rho}| \ll 1$ we can use the binomial expansion to approximate the density term as

$$\frac{1}{\bar{\rho} + \rho'} = \frac{1}{\bar{\rho}}\left(1 + \frac{\rho'}{\bar{\rho}}\right)^{-1} \simeq \frac{1}{\bar{\rho}}\left(1 - \frac{\rho'}{\bar{\rho}}\right)$$

Neglecting products of the perturbation quantities and noting that the basic state fields are constants, we obtain the linear perturbation equations[1]

$$\left(\frac{\partial}{\partial t} + \bar{u}\frac{\partial}{\partial x}\right)u' + \frac{1}{\bar{\rho}}\frac{\partial p'}{\partial x} = 0 \qquad (9.11)$$

$$\left(\frac{\partial}{\partial t} + \bar{u}\frac{\partial}{\partial x}\right)p' + \gamma\bar{p}\frac{\partial u'}{\partial x} = 0 \qquad (9.12)$$

Eliminating u' by operating on (9.12) with $(\partial/\partial t + \bar{u}\,\partial/\partial x)$ and substituting from (9.11) we get

$$\left(\frac{\partial}{\partial t} + \bar{u}\frac{\partial}{\partial x}\right)^2 p' - \frac{\gamma\bar{p}}{\rho}\frac{\partial^2 p'}{\partial x^2} = 0 \qquad (9.13)$$

Comparing with (9.1) we see that (9.13) is a form of the wave equation Therefore, following the discussion of the previous section, we assume a solution of the form

$$p' = Ae^{ik(x - ct)} \qquad (9.14)$$

where for brevity we omit the Re{ } notation but it is to be understood that

[1] It is not necessary that $|u'/\bar{u}| \ll 1$ for this linearization to be valid. In fact $\bar{u}$ may be zero, but (9.11) and (9.12) will still be approximately valid, provided u' and p' are sufficiently small so that the terms $u'\,\partial u'/\partial x$ and $u'\,\partial p'/\partial x$ are small compared to the terms retained in (9.11) and (9.12), respectively.

only the real part of (9.14) has physical significance. Substituting the assumed solution (9.14) into (9.13) we find that the phase speed c must satisfy

$$(-ikc + ik\bar{u})^2 - \frac{\gamma\bar{p}}{\bar{\rho}}(ik)^2 = 0$$

where we have canceled out the factor $Ae^{ik(x-ct)}$ which is common to both terms. Solving for c gives

$$c = \bar{u} \pm \left(\frac{\gamma\bar{p}}{\bar{\rho}}\right)^{1/2} = \bar{u} \pm (\gamma R\bar{T})^{1/2} \qquad (9.15)$$

Therefore (9.14) is a solution of (9.13) provided that the phase speed satisfies (9.15). According to (9.15) the speed of wave propagation relative to the zonal current is $(\gamma R\bar{T})^{1/2}$. This quantity is called the adiabatic speed of sound. The mean zonal velocity here plays only a role of *doppler shifting* the sound wave so that the frequency

$$\nu = kc = k\bar{u} \pm k(\gamma R\bar{T})^{1/2}$$

corresponding to a given wave number k, appears higher to an observer downstream from the source than to an upstream observer.

9.4 Gravity Waves: Vertical Stability

As a second example of pure wave motion we consider the transverse oscillations known as gravity waves. Gravity waves can only exist in a medium which is *stably stratified* so that a small fluid parcel which is displaced vertically from its equilibrium level will experience a net force which will tend to return it to the equilibrium level. A qualitative notion of the nature of gravity waves can be obtained by considering the dynamics of such a parcel under the assumption that the motion of the parcel does not affect the environment. This is the subject of the first subsection below. In the second subsection we treat an especially simple type of gravity wave which propagates along the interface between two homogeneous fluids of different density. Finally, in the third and fourth subsections we discuss some properties of internal atmospheric gravity waves.

9.4.1 VERTICAL OSCILLATIONS

The simplest approach to the concept of vertical stability is the *parcel method*. We consider a dry atmosphere which is in hydrostatic balance:

$$g = -\frac{1}{\bar{\rho}}\frac{\partial\bar{p}}{\partial z} \qquad (9.16)$$

where $\bar{p}$ and $\bar{\rho}$ are the pressure and density of the environment, respectively. If a small parcel of air is displaced vertically a short distance, the vertical acceleration of the parcel will be

$$\frac{dw}{dt} \equiv \frac{d^2z}{dt^2} = -g - \frac{1}{\rho}\frac{\partial p}{\partial z} \tag{9.17}$$

where p and ρ are the pressure and density of the parcel. In the parcel method it is assumed that the pressure of the parcel and environment remain equal during the displacement: $p = \bar{p}$. This condition must be true if the parcel is to leave the environment undisturbed. Thus, pressure can be eliminated between (9.16) and (9.17) to give

$$\frac{d^2z}{dt^2} = g\left(\frac{\bar{\rho} - \rho}{\rho}\right)$$

Or, in terms of potential temperature, using Poisson's equation (1.18) with the condition $p = \bar{p}$ we get

$$\frac{d^2z}{dt^2} = g\left(\frac{\theta - \bar{\theta}}{\bar{\theta}}\right) \tag{9.18}$$

If the parcel is initially at level $z = 0$ where the potential temperature is θ_0, then for a small displacement δz we can represent the environmental potential temperature as

$$\bar{\theta}(\delta z) \simeq \theta_0 + \frac{d\bar{\theta}}{dz}\delta z$$

If the parcel displacement is adiabatic, the potential temperature of the parcel is conserved: $\theta(\delta z) = \theta_0$. Thus, (9.18) becomes

$$\frac{d^2(\delta z)}{dt^2} = -N^2\delta z \tag{9.19}$$

where

$$N^2 = \frac{g}{\bar{\theta}}\frac{d\bar{\theta}}{dz}$$

is a measure of the static stability of the environment. Equation (9.19) has a general solution of the form

$$\delta z = Ae^{iNt}$$

Therefore, if $N > 0$ the parcel will oscillate about its initial level with a period

$$\tau = 2\pi/N \tag{9.20}$$

This period is called the period of a *buoyancy oscillation*, and the corresponding frequency N is often referred to as the Brunt–Vaisalla frequency.

In the case of $N = 0$, examination of (9.19) indicates that no accelerating force will exist and the parcel will be in neutral equilibrium at its new level. On the other hand, if $N^2 < 0$ (potential temperature decreasing with height) the displacement will increase exponentially in time. We thus arrive at the familiar gravitational or static stability criteria for dry air:

$$\frac{d\theta}{dz} \begin{cases} > 0 & \text{stable} \\ = 0 & \text{neutral} \\ < 0 & \text{unstable} \end{cases}$$

On the synoptic scale the atmosphere is always stably stratified because any unstable regions which develop are quickly stabilized by convective overturning. For a moist atmosphere, the situation is more complicated and discussion of that situation will be deferred until Chapter 12.

9.4.2 SHALLOW WATER WAVES

Before discussing the propagation of gravity waves in a continuously stratified atmosphere, it is worthwhile to examine the basic nature of gravity waves by considering the simpler problem of gravity waves propagating along the interface between two homogeneous incompressible fluids of differing density. The assumption of incompressibility is sufficient to exclude sound waves from the system, and we can thus isolate the gravity waves:

We consider a two-layer fluid system as shown in Fig. 9.2. If the density of

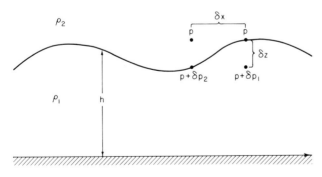

Fig. 9.2 A two-layer fluid system.

the lower layer ρ_1 is greater than the density of the upper layer ρ_2, the system is stably stratified. Since both ρ_1 and ρ_2 are constants, the horizontal pressure gradient in each layer is independent of height if the pressure is hydrostatic.

This may be verified by differentiating the hydrostatic approximation with respect to x

$$\frac{\partial}{\partial z}\left(\frac{\partial p}{\partial x}\right) = -\frac{\partial \rho}{\partial x} g = 0$$

For simplicity, we assume that there is no horizontal pressure gradient in the upper layer. The pressure gradient in the lower layer can be obtained by vertical integration of the hydrostatic equation. For the points shown in Fig. 9.2, we find

$$p + \delta p_1 = p + \rho_1 g \, \delta z = p + \rho_1 g \frac{\partial h}{\partial x} \delta x$$

$$p + \delta p_2 = p + \rho_2 g \, \delta z = p + \rho_2 g \frac{\partial h}{\partial x} \delta x$$

where $\partial h/\partial x$ is the slope of the interface. Taking the limit $\delta x \to 0$, we obtain the pressure gradient in the lower layer

$$\lim_{\delta x \to 0}\left[\frac{(p + \delta p_1) - (p + \delta p_2)}{\delta x}\right] = g \, \Delta\rho \frac{\partial h}{\partial x}$$

where $\Delta\rho = \rho_1 - \rho_2$.

We assume that the motion is two dimensional in the x, z plane. The x momentum equation for the lower layer is then

$$\frac{\partial u}{\partial t} + u \frac{\partial u}{\partial x} = -\frac{g \, \Delta\rho}{\rho_1} \frac{\partial h}{\partial x} \tag{9.21}$$

while the continuity equation is

$$\frac{\partial u}{\partial x} + \frac{\partial w}{\partial z} = 0 \tag{9.22}$$

Now since the pressure gradient in (9.21) is independent of z, u will also be independent of z. Thus, (9.22) can be integrated vertically from the lower boundary $z = 0$ to the interface $z = h$ to yield

$$w(h) - w(0) = -h\frac{\partial u}{\partial x}$$

But $w(h)$ is just the rate at which the interface height is changing

$$w(h) = \frac{dh}{dt} = \frac{\partial h}{\partial t} + u \frac{\partial h}{\partial x}$$

and $w(0) = 0$ for a flat lower boundary. Hence the vertically integrated continuity equation can be written

$$\frac{\partial h}{\partial t} + \frac{\partial}{\partial x}(hu) = 0 \tag{9.23}$$

Equations (9.21) and (9.23) are a closed set in the variables u and h. We now apply the perturbation technique by letting

$$u = \bar{u} + u', \qquad h = H + h'$$

where $\bar{u}$ as before is a constant basic state zonal velocity and H is the mean depth of the lower layer. The perturbation forms of (9.21) and (9.23) are then

$$\frac{\partial u'}{\partial t} + \bar{u}\frac{\partial u'}{\partial x} + \frac{g\,\Delta\rho}{\rho_1}\frac{\partial h'}{\partial x} = 0 \tag{9.24}$$

$$\frac{\partial h'}{\partial t} + \bar{u}\frac{\partial h'}{\partial x} + H\frac{\partial u'}{\partial x} = 0 \tag{9.25}$$

where we assume that $H \gg |h'|$ so that products of the perturbation variables can be neglected.

Eliminating u' between (9.24) and (9.25) we get

$$\left(\frac{\partial}{\partial t} + \bar{u}\frac{\partial}{\partial x}\right)^2 h' - \frac{gH\,\Delta\rho}{\rho_1}\frac{\partial^2 h}{\partial x^2} = 0 \tag{9.26}$$

which is a wave equation similar in form to (9.13). We again assume a wave-type solution by letting

$$h' = Ae^{ik(x - ct)}$$

Substituting into (9.26) we find that the assumed solution satisfies the equation provided that

$$c = \bar{u} \pm \left(\frac{gH\,\Delta\rho}{\rho_1}\right)^{1/2} \tag{9.27}$$

If the upper and lower layers are air and water, respectively, then $\Delta\rho \simeq \rho_1$ and the phase speed formula simplifies to

$$c = \bar{u} \pm \sqrt{gH}$$

The quantity $\sqrt{gH}$ is called the *shallow water* wave speed. It is a valid approximation only for waves whose wavelengths are much greater than the depth of the fluid. This restriction is necessary in order that the vertical velocities be small enough so that the hydrostatic approximation is valid. If the depth of the ocean is taken as 4 km, the shallow water gravity wave speed is $\simeq 200$ m sec^{-1}.

Thus, long waves on the ocean surface travel very rapidly. It should be emphasized again that this theory applies only to waves of wavelength much greater than H. Such long waves are not ordinarily excited by the wind stresses, but may be produced by very large-scale disturbances such as earthquakes.

Gravity waves may also occur at interfaces within the ocean where there is a very sharp density gradient (diffusion will always prevent formation of a true density discontinuity). In particular, the surface water is separated from the deep water by a narrow region of sharp density contrast called the *thermocline*. If the thermocline is regarded as an interface across which the density changes by an amount $\Delta\rho \simeq 0.01$, then from (9.27) it is clear that the wave speed for waves traveling along the thermocline will be only one-tenth of the surface wave speed for a system of the same depth.

9.4.3 INTERNAL GRAVITY WAVES

In a fluid, such as the ocean, which is bounded both above and below, gravity waves propagate primarily in the horizontal plane since vertically traveling waves would be reflected from the boundaries to form standing waves. However, in a fluid which has no upper boundary, such as the atmosphere, internal gravity waves may propagate *vertically* as well as horizontally. Although vertically propagating internal gravity waves are not important for short-range synoptic scale forecasting (and indeed are nonexistent in the filtered quasi-geostrophic models), they are responsible for the occurrence of mountain *lee waves*. They also are believed to be an important mechanism for transporting energy and momentum to high levels, and are often associated with the formation of clear air turbulence (CAT).

To analyze the properties of internal atmospheric gravity waves, it is convenient to write the perturbation equations in a coordinate system in which the vertical coordinate is proportional to the log of pressure. In this system, as in the usual isobaric system, density does not explicitly appear. However, unlike the case for the isobaric system, the static stability parameter in the log-pressure system is nearly constant with respect to height in the troposphere. The log-pressure system thus combines some of the best features of both the isobaric coordinates and height coordinates.

In the log-pressure system, the vertical coordinate is defined as

$$z^* \equiv -H \ln \frac{p}{p_0} \qquad (9.28)$$

where p_0 is a standard reference pressure (usually taken to be 1000 mb) and $H \equiv RT_0/g$ is a constant *scale height*. In Problem 4 at the end of this chapter the student is asked to show that H defined in this manner is just the height in which the pressure of an isothermal atmosphere at temperature T_0 decays to

e^{-1} of its surface value. Thus for an isothermal atmosphere z^* as defined in (9.28) is exactly equal to height. For an atmosphere with variable temperature, z^* will be only approximately equivalent to the actual height. The vertical velocity in this coordinate system is defined as

$$w^* \equiv \frac{dz^*}{dt} \qquad (9.29)$$

The horizontal momentum equation in the log-pressure system is the same as that in the isobaric system,

$$\frac{d\mathbf{V}}{dt} + f\mathbf{K} \times \mathbf{V} = -\nabla\Phi \qquad (9.30)$$

However, the operator d/dt is now defined as

$$\frac{d}{dt} \equiv \frac{\partial}{\partial t} + \mathbf{V} \cdot \nabla + w^* \frac{\partial}{\partial z^*}$$

The hydrostatic equation

$$\frac{\partial\Phi}{\partial p} = -\alpha$$

may be transformed to the log-pressure system by eliminating α with the ideal gas law to get

$$\frac{\partial\Phi}{\partial \ln p} = -RT$$

or upon dividing through by $-H$

$$\frac{\partial\Phi}{dz^*} = \frac{RT}{H} \qquad (9.31)$$

The continuity equation may also be conveniently obtained by transforming from the isobaric coordinate form

$$\frac{\partial u}{\partial x} + \frac{\partial v}{\partial y} + \frac{\partial \omega}{\partial p} = 0$$

From (9.28) and (9.29) we have

$$w^* = -\frac{H}{p}\frac{dp}{dt} = -\frac{H\omega}{p}$$

from which we see that

$$\frac{\partial\omega}{\partial p} = -\frac{\partial}{\partial p}\left(\frac{pw^*}{H}\right) = \frac{\partial w^*}{\partial z^*} - \frac{w^*}{H}$$

Thus, in log-pressure coordinates the continuity equation becomes simply

$$\frac{\partial u}{\partial x} + \frac{\partial v}{\partial y} + \frac{\partial w^*}{\partial z^*} - \frac{w^*}{H} = 0 \tag{9.32}$$

It is left as a problem for the student to show that the first law of thermodynamics (1.17), which can be expressed in the form

$$c_p \frac{dT}{dt} - \alpha \frac{dp}{dt} = \dot{H} \tag{9.33}$$

where $\dot{H}$ is the heating rate per unit mass, has the following form in the log-pressure system:

$$\left(\frac{\partial}{\partial t} + \mathbf{V} \cdot \mathbf{V}\right)T + w^*\Gamma = \frac{1}{c_p}\dot{H} \tag{9.34}$$

where

$$\Gamma = \frac{\partial T}{\partial z^*} + \frac{RT}{c_p H} = \frac{T}{\theta}\frac{\partial \theta}{\partial z^*}$$

is the static stability parameter. For the troposphere, $\Gamma \simeq 3°C \text{ km}^{-1}$ is nearly constant, whereas the stability parameter

$$\sigma \equiv -\frac{\alpha}{\theta}\frac{\partial \theta}{\partial p}$$

of the isobaric system varies substantially with height. This near constancy of Γ is one of the chief advantages of the log-pressure coordinates for analysis of internal gravity waves.

We now assume that the motion is restricted to the x,z^* plane ($v = 0$) and is independent of y. The perturbation equations of motion, continuity, and thermodynamic energy in log-pressure coordinates then become

$$\frac{\partial u'}{\partial t} + \bar{u}\frac{\partial u'}{\partial x} + \frac{\partial \Phi'}{\partial x} = 0 \tag{9.35}$$

$$\frac{\partial u'}{\partial x} + \frac{\partial w^{*'}}{\partial z^*} - \frac{w^{*'}}{H} = 0 \tag{9.36}$$

$$\frac{\partial}{\partial t}\left(\frac{\partial \Phi'}{\partial z^*}\right) + \bar{u}\frac{\partial}{\partial x}\left(\frac{\partial \Phi'}{\partial z^*}\right) + w^{*'}S = 0 \tag{9.37}$$

Here $\bar{u}$ is a constant zonal current, $S \equiv R\Gamma/H$ is assumed to be constant, and the motion is assumed to be adiabatic.

Combining (9.36) and (9.37) to eliminate $w^{*'}$, we get

$$\left(\frac{\partial}{\partial t} + \bar{u}\frac{\partial}{\partial x}\right)\left(\frac{\partial^2 \Phi'}{\partial z^{*2}} - \frac{1}{H}\frac{\partial \Phi'}{\partial z^*}\right) - S\frac{\partial u'}{\partial x} = 0 \tag{9.38}$$

And eliminating u' between (9.35) and (9.38) gives

$$\left(\frac{\partial}{\partial t} + \bar{u}\frac{\partial}{\partial x}\right)^2\left(\frac{\partial^2 \Phi'}{\partial z^{*2}} - \frac{1}{H}\frac{\partial \Phi'}{\partial z^*}\right) + S\frac{\partial^2 \Phi'}{\partial x^2} = 0 \tag{9.39}$$

Equation (9.39) is a partial differential equation with independent variables x, z^*, and t. We may separate the vertical and horizontal dependencies by again assuming a solution in the form of a zonally propagating wave:

$$\Phi' = A(z^*)e^{i(kx - vt)} \tag{9.40}$$

where v is the angular frequency of oscillation. Substituting the assumed solution (9.40) into (9.39), we find that the coefficient $A(z^*)$ must satisfy

$$\frac{d^2 A}{dz^{*2}} - \frac{1}{H}\frac{dA}{dz^*} + \frac{S}{(\bar{u} - v/k)^2}A = 0 \tag{9.41}$$

The general solution of (9.41) has the form

$$A(z^*) = e^{z^*/2H}(C_1 e^{-i\lambda z^*} + C_2 e^{i\lambda z^*})$$

where

$$\lambda = \left[\frac{S}{(\bar{u} - v/k)^2} - \frac{1}{4H^2}\right]^{1/2} \tag{9.42}$$

Thus λ is the vertical wave number. The geopotential perturbation may be written

$$\Phi' = [C_1 e^{i(kx - \lambda z^* - vt)} + C_2 e^{i(kx + \lambda z^* - vt)}]e^{z^*/2H} \tag{9.43}$$

where C_1 and C_2 are constants which must be determined by the boundary conditions.

If we assume that k, λ, and v are all positive, then the first term on the right in (9.43) corresponds to a wave whose phase speed has a downward component, whereas the second term corresponds to a wave whose phase speed has an upward component. The phase relationships for a wave whose phase velocity is downward and eastward are shown in Fig. 9.3. In this case the constant phase lines are given by

$$kx - \lambda z^* - vt = \text{const}$$

The slope of the phase lines is thus k/λ. The horizontal component of the phase velocity is $c = v/k$ and the vertical component of the phase velocity is v/λ. For a typical tropospheric value of $S \simeq 10^{-4}$ cgs, a wave of vertical wavelength ~ 2 km will, according to (9.42), have a horizontal phase speed of only ~ 3 m sec^{-1}.

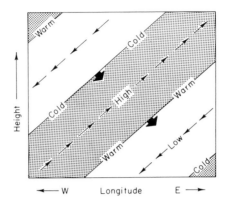

Fig. 9.3   Idealized cross section showing phases of the geopotential, temperature, and velocity perturbations for an internal gravity wave. Thin arrows indicate phase of perturbation velocity field, blunt arrows indicate the phase velocity. (After Wallace and Kousky, 1968.)

Substituting the solution (9.43) for the perturbation geopotential into (9.38) and (9.37), we can obtain the corresponding fields of u' and $w*'$. These are indicated schematically in Fig. 9.3. It is clear from the figure that u' and $w*'$ are positively correlated, so that the motion is upward when the perturbation zonal velocity is positive, and downward when it is negative. Thus, an internal gravity wave whose phase speed is downward and eastward transports positive zonal momentum *upward*. It can be shown from an analysis of the energy equations that such waves also transport energy upward. Thus, downward phase propagation implies upward energy and momentum propagation.

9.4.4   GROUP VELOCITY

The relationship between phase propagation and energy propagation in internal gravity waves can be further elucidated by considering a wave disturbance composed of Fourier components whose zonal wave numbers lie between $k + \Delta k$ and $k - \Delta k$. In fact any *transient* disturbance must have its energy distributed over a number of Fourier components since a pure harmonic with a single wave number has constant amplitude for all x space. On the other hand, since the individual components of a wave group alternately reinforce and cancel each other, the energy of the group will be concentrated in a limited region, as illustrated in Fig. 9.4. If the phase velocity of the waves in a group does not depend on wavelength, the group will propagate with the phase velocity without changing shape. However, from (9.42) we see that the phase velocity of an internal gravity wave depends on the vertical wavelength. Dependence of the phase velocity of waves on wavelength is called *dispersion*.

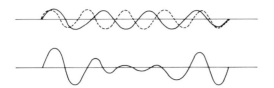

Fig. 9.4 Wave group composed of two sinusoidal components of slightly different wavelengths.

When waves are dispersive, the speed of the wave group is generally different from the average phase speed of the individual Fourier components. Hence, as shown in Fig. 9.5, individual wave components may move through the group as the group propagates along. Furthermore, the group generally broadens in the course of time, that is, the energy is *dispersed.*

An expression for the *group velocity* which is the velocity at which the observable disturbance, and hence the energy, propagates can be derived as follows: We consider two horizontally propagating waves of equal amplitude but slightly different wavelengths with wave numbers and frequencies differing by $2 \Delta k$ and $2 \Delta v$, respectively. The amplitude of the total disturbance is thus

$$A = e^{i[(k + \Delta k)x - (v + \Delta v)t]} + e^{i[(k - \Delta k)x - (v - \Delta v)t]}$$

Rearranging terms we get

$$A = [e^{i(\Delta kx - \Delta vt)} + e^{-i(\Delta kx - \Delta vt)}]e^{i(kx - vt)}$$

or

$$A = 2 \cos(\Delta kx - \Delta vt)e^{i(kx - vt)} \tag{9.44}$$

The disturbance (9.44) is the product of a high-frequency *carrier wave* of wavelength $2\pi/k$ whose phase velocity is the average for the two Fourier components, and a low frequency *envelope* of wavelength $2\pi/\Delta k$ which travels at the

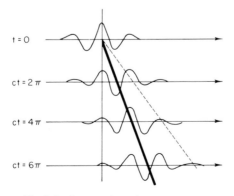

Fig. 9.5 Propagation of a wave group.

speed $\Delta v/\Delta k$. Thus, in the limit as $\Delta k \to 0$, the horizontal velocity of the envelope, or *group velocity*, is just

$$c_g = \frac{\partial v}{\partial k}$$

By a similar argument it can be shown that the vertical component of the group velocity is $\partial v/\partial \lambda$.

Applying this expression to the case of internal gravity waves we find from (9.42) that if $\bar{u} = 0$, and $\lambda H \gg 1$

$$\frac{\partial v}{\partial \lambda} = -\frac{kS^{1/2}}{\lambda^2} = -\frac{v}{\lambda} \tag{9.45}$$

Hence, the vertical component of group velocity is equal and opposite to the vertical phase speed. As a consequence an upward traveling group will have downward phase propagation. Since energy travels at the group velocity, we have confirmed the assertion at the end of the last section that an internal wave with a downward component of phase velocity carries energy upward.

9.4.5 LEE WAVES

When air is forced to flow over a mountain under statically stable conditions, individual air parcels are displaced from their equilibrium levels and will thus undergo buoyancy oscillations as they move downstream of the mountain. In this manner, as shown in Fig. 9.6, an internal gravity wave system

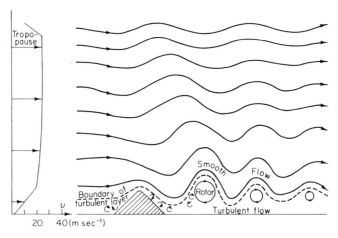

Fig. 9.6 Schematic diagram showing streamlines based on observed lee waves. The upstream velocity profile is indicated on the left. (After Gerbier and Berenger, 1961.)

is excited in the lee of the mountain stationary with respect to the ground. If the vertical motion associated with these *lee waves* is strong enough, and the air is moist enough, condensation may occur in the updraft portions of the oscillations giving rise to wave clouds. Such clouds are a common occurrence to the east of large north–south oriented mountain barriers such as the Rockies.

Since lee waves are stationary with respect to the ground, $c = 0$, and from (9.42) we find that[2]

$$\lambda = \left[\frac{S}{\bar{u}^2} - \frac{1}{4H^2}\right]^{1/2} \simeq \frac{S^{1/2}}{\bar{u}}$$

Thus, the vertical wavelength of the gravity wave excited by zonal flow over a mountain is proportional to the zonal wind speed and inversely proportional to the square root of the stability. When viewed in a coordinate system moving at the speed of the mean zonal wind, constant phase lines of lee waves setup by westerly flow appear to progress upstream toward the west. Since the energy source for lee waves is at the ground, such waves must transport energy upward. Hence, the phase velocity relative to the mean zonal flow must have a downward component. Consequently, the lines of constant phase must tilt westward with height as shown in Fig. 9.6.

9.5 Rossby Waves

The wave type which is of most importance for large-scale atmospheric flow processes is the *Rossby wave*. Rossby waves owe their existence to the variation of the Coriolis force with latitude, the so-called β effect. In order to isolate the Rossby wave from other types of atmospheric disturbances, we refer the motion to the isobaric coordinate system, which eliminates vertically propagating sound waves, and assume that $\omega \equiv 0$, which eliminates the horizontal acoustic waves as well as gravity waves. The x and y components of the momentum equation and the continuity condition can then be written

$$\left(\frac{\partial}{\partial t} + u\frac{\partial}{\partial x} + v\frac{\partial}{\partial y}\right)u - fv = -\frac{\partial\Phi}{\partial x} \tag{9.46}$$

$$\left(\frac{\partial}{\partial t} + u\frac{\partial}{\partial x} + v\frac{\partial}{\partial y}\right)v + fu = -\frac{\partial\Phi}{\partial y} \tag{9.47}$$

$$\frac{\partial u}{\partial x} - \frac{\partial v}{\partial y} = 0 \tag{9.48}$$

---

[2] Lee waves depart significantly from hydrostatic equilibrium, but (9.42) is nevertheless generally qualitatively valid.

Forming the vorticity equation from (9.46) and (9.47) and noting from (9.48) that the divergence term is zero, we obtain

$$\left(\frac{\partial}{\partial t} + u \frac{\partial}{\partial x} + v \frac{\partial}{\partial y}\right)\zeta + v \frac{df}{dy} = 0 \tag{9.49}$$

which states that the absolute vorticity is conserved following the horizontal motion in isobaric coordinates. Equation (9.49) is, of course, just the barotropic vorticity equation discussed in Section 8.4.1.

We now assume that the motion consists of a basic state zonal velocity plus a small horizontal perturbation:

$$u = \bar{u} + u', \qquad v = v', \qquad \zeta = \zeta'$$

We define a perturbation streamfunction ψ according to

$$u' = -\frac{\partial \psi}{\partial y}, \qquad v' = \frac{\partial \psi}{\partial x}$$

from which $\zeta' = \nabla^2 \psi$. The perturbation form of (9.49) is then

$$\left(\frac{\partial}{\partial t} + \bar{u} \frac{\partial}{\partial x}\right)\nabla^2 \psi + \beta \frac{\partial \psi}{\partial x} = 0 \tag{9.50}$$

where we now assume that $\beta \equiv df/dy$ is a constant, and as usual we have neglected terms involving the products of perturbation quantities. We assume a solution

$$\psi = A e^{ik(x - ct)} \cos my \tag{9.51}$$

Thus, k is the zonal wavelength, c is the zonal phase speed, and m is the latitudinal wavenumber. Substituting (9.51) into (9.50) gives

$$(-ikc + ik\bar{u})(-k^2 - m^2) + ik\beta = 0$$

In order that the assumed solution satisfy (9.50), the phase speed must be given by

$$c = \bar{u} - \frac{\beta}{(k^2 + m^2)} \tag{9.52}$$

Thus, the Rossby wave propagates *westward* relative to the mean zonal flow. Furthermore the Rossby wave speed depends on the zonal and meridional wave numbers. Therefore, Rossby waves are dispersive waves whose phase speeds increase with wavelength. This result is consistent with the discussion in Section 7.3.1 in which we showed that the advection of planetary vorticity, which tends to make disturbances retrogress, increasingly dominates over relative vorticity advection as the wavelength of a disturbance increases.

Equation (9.52) merely provides a quantitative measure of this effect in cases where the disturbance is small enough so that perturbation theory is applicable. For a typical midlatitude synoptic disturbance with zonal wavelength ~ 6000 km and latitudinal width ~ 3000 km, the Rossby wave speed relative to the zonal flow calculated from (9.52) is approximately -6 m sec^{-1}. Thus, synoptic scale Rossby waves move quite slowly.

It is possible to carry out a less restrictive analysis using the perturbation form of the full primitive equations. The results of such an analysis are mathematically complicated but qualitatively not much different from the pure wave cases discussed here. It turns out that the free oscillations allowed in a hydrostatic gravitationally stable atmosphere consist of eastward and westard moving gravity waves which are slightly modified by the rotation of the earth, and westward moving Rossby waves which are slightly modified by gravitational stability. These free oscillations are the *normal modes* of oscillation of the atmosphere. As such, they are continually excited by the various forces acting on the atmosphere. The planetary scale free oscillations, although they can be detected by careful observational studies, appear to have rather weak amplitudes. Presumably this is because the forcing is quite weak at the large phase speeds charateristic of these long waves.

Problems

1. Verify that $\psi = f(x - ct)$ is a solution of Eq. (9.1) for any arbitrary continuous profile $f(x - ct)$. *Hint*: Let $x - ct = x'$ and differentiate f using the chain rule.

2. Show that the Fourier component

$$F(x) = \text{Re}\{Ce^{imx}\}$$

can be written as

$$F(x) = |C| \cos m(x + x_0)$$

where $x_0 = m^{-1} \sin^{-1}[C_i/|C|]$ and C_i stands for the imaginary part of C.

3. In the study of atmospheric wave motions it is often necessary to consider the possibility of amplifying or decaying waves. In such a case we might assume that a solution has the form

$$\psi = A^{\alpha t} \cos(kx - vt - kx_0)$$

where A is the initial amplitude, α the amplification factor, and x_0 the initial phase. Show that this expression can be written more concisely as

$$\psi = \text{Re}\{Be^{ik(x - ct)}\}$$

where both B and c are complex constants. Determine the real and imaginary parts of B and c in terms of A, α, k, v, and x_0.

4. Show with the aid of the hydrostatic approximation and the ideal gas law that the scale height $H \equiv RT/g$ is the height in which pressure and density decrease by a factor of e in an isothermal atmosphere.

5. Starting with the first law of thermodynamics for an ideal gas in the form (9.33), derive the log-pressure coordinate form (9.34).

6. Derive the Rossby wave speed for a homogeneous incompressible ocean of depth h. Assume a motionless basic state and small perturbations which depend only on x and t:

$$u = u'(x, t), \qquad v = v'(x, t), \qquad h = H + h'(x, t)$$

where H is the mean depth of the ocean. With the aid of the continuity equation for a homogeneous layer (9.25) and the geostrophic wind relationship

$$v' = \frac{g}{f}\frac{h'}{x}$$

show that the perturbation vorticity equation can be written

$$\frac{\partial}{\partial t}\left(\frac{\partial^2}{\partial x^2} - \frac{f^2}{gH}\right)h' + \beta\frac{\partial h'}{\partial x} = 0$$

and that $h' = Ae^{ik(x-ct)}$ is a solution provided that

$$c = -\frac{\beta}{(k^2 + f^2/gH)}$$

If the ocean is 4 km deep, what is the Rossby wave speed at latitude $45°$ for a wave of 10,000-km zonal wavelength?

7. In Section 5.3 we showed that for a homogeneous incompressible fluid a decrease in depth with latitude has the same dynamic effect as a latitudinal dependence of the Coriolis parameter. Thus, Rossby-type waves can be produced in a rotating cylindrical vessel if the depth of the fluid is dependent on the radial coordinate. To determine the Rossby wave speed formula for this so-called "equivalent β effect," we assume that the flow is confined between rigid lids in an annular region whose distance from the axis of rotation is large enough so that the curvature terms in the equations can be neglected. We can then refer the motion to cartesian coordinates with x directed azimuthally and y directed toward the axis of rotation. If the system is rotating at angular velocity Ω and the depth is linearly dependent on y,

$$H(y) = H_0 - \gamma y$$

show that the perturbation continuity equation can be written

$$H_0 \left(\frac{\partial u'}{\partial x} + \frac{\partial v'}{\partial y} \right) - \gamma v' = 0$$

and that the perturbation quasi-geostrophic vorticity is thus

$$\frac{\partial}{\partial t} \nabla^2 h' + \beta \frac{\partial h'}{\partial x} = 0$$

where

$$\beta = \frac{2\Omega}{H_0} \gamma$$

Hint: Assume that the velocity field is geostrophic except in the divergence term. What is the Rossby wave speed in this situation for waves of wavelength 100 cm in both the x and y directions if $\Omega = 1 \ \text{sec}^{-1}$, $H_0 = 20$ cm, and $\gamma = 0.05$?

8. Show by scaling arguments that if the horizontal wavelength is much greater than the depth of the fluid, two-dimensional surface gravity waves will be hydrostatic so that the " shallow water " approximation applies.

9. Derive an expression for the group velocity of a barotropic Rossby wave and compare with the phase speed.

Suggested References

Hildebrand, *Advanced Calculus for Applications*, is one of many standard textbooks which discusses the mathematical techniques used in this chapter, including the representation of functions in Fourier series and the general properties of the wave equation.

Eckart, *Hydrodynamics of Oceans and Atmospheres*, contains an advanced treatment of acoustic and gravity wave motions in stratified rotating fluids.

Chapman and Lindzen, *Atmospheric Tides*, thoroughly covers both observational and theoretical aspects of a class of atmospheric motions in which analysis by the linear perturbation method has proved particularly successful.

Scorer, *Natural Aerodynamics*, contains an excellent qualitative discussion on many aspects of waves generated by barriers such as lee waves.

Chapter

10 | The Origin and Motion of Midlatitude Synoptic Systems

In Chapter 7 we discussed the observed structure of midlatitude synoptic disturbances and showed that simple diagnostic relationships based on the quasi-geostrophic system can qualitatively account for the observed relationships among the pressure, temperature, and velocity fields. In particular, we found that horizontal temperature advection plays an essential role in the amplification of quasi-geostrophic disturbances. However, such diagnostic studies, although useful for interpreting the structure of baroclinic systems, are in themselves inadequate to determine the *origin* of these disturbances.

The presently accepted view is that baroclinic synoptic scale disturbances in the middle latitudes are initiated as the result of a *hydrodynamic instability* of the basic zonal current with respect to small perturbations of the flow. In this chapter we will examine this instability hypothesis for the origin of baroclinic waves and the energy conversions involved in the development of such waves. We will also consider briefly the development of fronts in association with synoptic disturbances.

10.1 Hydrodynamic Instability

The concept of hydrodynamic instability may be qualitatively understood by considering the motion of an individual fluid parcel in a steady zonal current. In Chapter 9 the parcel method was used to derive the criterion for static stability by examining the conditions under which a parcel displaced vertically without disturbing its surroundings would either return to its original position or be accelerated further from the initial level. The parcel method can be generalized by assuming that the parcel displacement is in an arbitrary direction. If the basic state is stably stratified and the zonal current has horizontal and vertical shear, the analysis becomes quite complicated. However, if the parcel is displaced horizontally across the basic flow, then the buoyancy force plays no role. This special case, called *inertial stability*, can be analyzed very simply if the basic flow is assumed to be geostrophic.

If we designate the basic state flow by

$$u_g = -\frac{1}{f}\frac{\partial \Phi}{\partial y}$$

and assume that the parcel displacement does not perturb the pressure field, the approximate equations of motion become

$$\frac{du}{dt} = fv = f\frac{dy}{dt} \tag{10.1}$$

$$\frac{dv}{dt} = f(u_g - u) \tag{10.2}$$

We consider a parcel which is moving with the geostrophic basic state motion at a position $y = y_0$. If the parcel is displaced across stream by a distance y, we can obtain its new zonal velocity by integrating (10.1)

$$u(y_0 + \delta y) = u_g(y_0) + f\,\delta y \tag{10.3}$$

The geostrophic wind at $y_0 + \delta y$ can be approximated as

$$u_g(y_0 + \delta y) = u_g(y_0) + \frac{\partial u_g}{\partial y}\,\delta y \tag{10.4}$$

Substituting from (10.3) and (10.4) into (10.2) we obtain

$$\frac{dv}{dt} = \frac{d^2(\delta y)}{dt^2} = -f\left(f - \frac{\partial u_g}{\partial y}\right)\delta y \tag{10.5}$$

This equation is mathematically of the same form as (9.19), the equation for the motion of a vertically displaced particle in a stable atmosphere. Depending on the sign of the coefficient on the right-hand side in (10.5) the parcel will either be forced to return to its original position or will accelerate further from that position. In the Northern Hemisphere where f is positive the inertial stability condition thus becomes

$$f - \frac{\partial u_g}{\partial y} \begin{array}{l} > \\ = \\ < \end{array} \begin{cases} 0 & \text{stable} \\ 0 & \text{neutral} \\ 0 & \text{unstable} \end{cases} \tag{10.6}$$

Since $f - \partial u_g/\partial y$ is the absolute vorticity of the basic flow, the inertial stability condition is simply that the absolute vorticity be positive. Observations indicate that on the synoptic scale the absolute vorticity is nearly always positive. The occurrence of a negative absolute vorticity over any large area would be expected to trigger immediately inertially unstable motions which would mix the fluid laterally and reduce the shear until the absolute vorticity was again positive. This mechanism is called *inertial instability* since when viewed in an absolute reference frame the instability results from an imbalance between the pressure gradient and centrifugal (that is, inertial) forces for a parcel displaced radially in an axisymmetric vortex.

Inertial and static instability are merely two of many possible types of hydrodynamic instability. In general, a basic flow subject to arbitrary perturbations may be subject to a variety of modes of instability which depend on the horizontal and vertical shear, the static stability, the variation of the Coriolis parameter, the influence of friction, etc. In only a few cases can the simple parcel method give satisfactory stability criteria. Usually a more rigorous approach is required in which a linearized version of the governing equations is analyzed to determine the conditions under which the solutions describe amplifying disturbances. As indicated in Problem 3 of Chapter 9, the usual approach is to assume a wave type solution of the form

$$e^{ik(x - ct)}$$

and to determine the conditions for which the phase velocity c has an imaginary part. This technique, which is called the *normal modes* method, will be applied in the next section to analyze the stability of a baroclinic current.

10.2 Baroclinic Instability: Cyclogenesis

The process of cyclogenesis will be regarded here as a manifestation of the amplification of an infinitesimal perturbation superposed on an unstable zonal current. For a perturbation to amplify, it is clear that the basic flow must give up potential and (or) kinetic energy to the perturbation. In baroclinic instability it turns out, as we shall show later, that the potential energy of the basic state flow is converted to potential and kinetic energy of the perturbation.

In this section we will derive the conditions for baroclinic instability using the two-level quasi-geostrophic model discussed in Chapter 8. This model certainly oversimplifies the vertical structure of baroclinic systems, but it does nevertheless contain the essential features necessary for a qualitative understanding of baroclinic instability.

We recall from (8.27)–(8.29) that the basic equations of the two-level model are

$$\frac{\partial}{\partial t} \nabla^2 \psi_1 + \mathbf{V}_1 \cdot \nabla(\nabla^2 \psi_1 + f) = \frac{f_0}{\Delta p} \omega_2 \qquad (10.7)$$

$$\frac{\partial}{\partial t} \nabla^2 \psi_3 + \mathbf{V}_3 \cdot \nabla(\nabla^2 \psi_3 + f) = -\frac{f_0}{\Delta p} \omega_2 \qquad (10.8)$$

$$\frac{\partial}{\partial t}(\psi_1 - \psi_3) + \mathbf{V}_2 \cdot \nabla(\psi_1 - \psi_3) = \frac{\sigma \Delta p}{f_0} \omega_2 \qquad (10.9)$$

where

$$\mathbf{V}_j = \mathbf{k} \times \nabla \psi_j \qquad \text{for} \quad j = 1, 2, 3$$

The arrangement of these variables in the vertical was shown in Fig. 8.4. To keep the analysis as simple as possible we now assume that the stream functions ψ_1 and ψ_3 consist of basic state parts which depend linearly on y alone, plus perturbations which depend only on x and t. Thus we let

$$\psi_1 = -U_1 y + \psi_1'(x, t)$$

$$\psi_3 = -U_3 y + \psi_3'(x, t) \qquad (10.10)$$

$$\omega_2 = \omega_2'(x, t)$$

The zonal velocities at levels 1 and 3 are then U_1 and U_3, respectively. Hence, the perturbation field has meridional and vertical velocity components only.

Substituting from (10.10) into (10.7)–(10.9) we obtain the perturbation equations

$$\left(\frac{\partial}{\partial t} + U_1 \frac{\partial}{\partial x}\right) \frac{\partial^2 \psi_1'}{\partial x^2} + \beta \frac{\partial \psi_1'}{\partial x} = \frac{f_0}{\Delta p} \omega_2'$$
$$(10.11)$$

$$\left(\frac{\partial}{\partial t} + U_3 \frac{\partial}{\partial x}\right) \frac{\partial^2 \psi_3'}{\partial x^2} + \beta \frac{\partial \psi_3'}{\partial x} = -\frac{f_0}{\Delta p} \omega_2'$$
$$(10.12)$$

$$\left(\frac{\partial}{\partial t} + \frac{U_1 + U_3}{2} \frac{\partial}{\partial x}\right)(\psi_1' - \psi_3') - \frac{U_1 - U_3}{2} \frac{\partial}{\partial x}(\psi_1' + \psi_3') = \frac{\sigma \Delta p}{f_0} \omega_2'$$
$$(10.13)$$

where we have used the β-plane approximation, $\beta \equiv df/dy$, and have linearly interpolated to express $\mathbf{V}_2$ in terms of ψ_1 and ψ_3. Equations (10.11)–(10.13) are a linear set in ψ_1', ψ_3', and ω_2'. As in Chapter 9 we assume wave-type solutions

$$\psi_1' = Ae^{ik(x-ct)}, \qquad \psi_3' = Be^{ik(x-ct)}, \qquad \omega_2' = Ce^{ik(x-ct)} \qquad (10.14)$$

Substituting these assumed solutions into (10.11)–(10.13) we find that the amplitude factors A, B, and C must satisfy the following set of simultaneous, homogeneous, linear algebraic equations:

$$ik[(c - U_1)k^2 + \beta]A - \frac{f_0}{\Delta p} C = 0 \qquad (10.15)$$

$$ik[(c - U_3)k^2 + \beta]B + \frac{f_0}{\Delta p} C = 0 \qquad (10.16)$$

$$-ik(c - U_3)A + ik(c - U_1)B - \frac{\sigma \Delta p}{f_0} = 0 \qquad (10.17)$$

Since this set is homogeneous, nontrivial solutions will exist only if the determinant of the coefficients of A, B, and C is zero. Thus we require that the phase speed c satisfy

$$\begin{vmatrix} ik[(c - U_1)k^2 + \beta] & 0 & -f_0/\Delta p \\ 0 & ik[(c - U_3)k^2 + \beta] & f_0/\Delta p \\ -ik(c - U_3) & ik(c - U_1) & -\sigma \Delta p/f_0 \end{vmatrix} = 0$$

Multiplying out the terms in the determinant we obtain a quadratic equation in c

$$(k^4 + 2\lambda^2 k^2)c^2 + [2\beta(k^2 + \lambda^2) - (U_1 + U_3)(k^4 + 2\lambda^2 k^2)]c$$
$$+ [k^4 U_1 U_3 + \beta^2 - (U_1 + U_3)(k^2 + \lambda^2)\beta + \lambda^2 k^2(U_3^2 + U_1^2)] = 0 \qquad (10.18)$$

where we have let $\lambda^2 \equiv f_0^2/(\sigma \Delta p^2)$. Alternatively, (10.18) could also have been obtained by eliminating any two of the variables A, B, or C between (10.15)–(10.17). Solving (10.18) for the phase speed we obtain

$$c = U_m - \frac{\beta(k^2 + \lambda^2)}{k^2(k^2 + 2\lambda^2)} \pm \delta^{1/2} \qquad (10.19)$$

where

$$\delta \equiv \frac{\beta^2 \lambda^4}{k^4(k^2 + 2\lambda^2)^2} - \frac{U_T^2(2\lambda^2 - k^2)}{(k^2 + 2\lambda^2)}$$

and

$$U_m \equiv \frac{U_1 + U_3}{2}, \qquad U_T \equiv \frac{U_1 - U_3}{2}$$

Thus U_m and U_T are, respectively, the vertically averaged zonal wind and the basic state thermal wind for the interval $\Delta p/2$.

We have now shown that (10.14) is a solution for the system (10.11)–(10.13) only if the phase speed satisfies (10.19). Equation (10.19) is a rather complicated expression. However, we can note immediately that if $\delta < 0$ the phase speed will have an imaginary part and the perturbations will amplify exponentially. Before discussing the general properties of (10.19) in detail, we will consider two special cases.

As the first special case, we consider a barotropic basic state. If we let $U_T = 0$ so that the basic state thermal wind vanishes the phase speeds

$$c_1 = U_m - \frac{\beta}{k^2} \tag{10.20}$$

and

$$c_2 = U_m - \frac{\beta}{(k^2 + 2\lambda^2)} \tag{10.21}$$

both are solutions. These are real quantities which correspond to the free (normal mode) oscillations for the two-level model with a barotropic basic state current. The phase speed c_1 is simply the dispersion relationship for a barotropic Rossby wave with no y dependence (see Section 9.5). Substituting the expression (10.20) in place of c in (10.15)–(10.17) we see that in this case $A = B$ and $C = 0$ so that the *perturbation is barotropic*. The expression (10.21), on the other hand, may be interpreted as the phase speed for an internal baroclinic Rossby wave. Note that c_2 is a dispersion relationship analogous to the Rossby wave speed for a homogeneous ocean with a free surface which was given in Problem 4 of Chapter 9. But, in the two-level model, the factor $2\lambda^2 \equiv 2f_0^2/\sigma \, \Delta p^2$ appears in the denominator in place of the f_0^2/gH for the oceanic case. In each of these cases there is vertical motion associated with the Rossby wave so that static stability modifies the wave speed. It is left as a problem for the reader to show that if c_2 is substituted into (10.15)–(10.17), the resulting fields of ψ_1 and ψ_3 are $180°$ out of phase so that the perturbation is baroclinic, although the basic state is barotropic. Furthermore, if $U_m > \beta/(k^2 + 2\lambda^2)$ so that the wave moves eastward, the ω_2 field leads the 250-mb geopotential field by $90°$ phase (that is, the maximum upward motion occurs east of the 250-mb trough). This vertical motion pattern is, of course, just what is required to keep the thickness field hydrostatic in the presence of the strong differential vorticity advection associated with the baroclinic Rossby mode. Hence, the present result is consistent with the diagnostic "omega" equation for the two-level model (8.34). Comparing (10.20) and (10.21) we see that the phase speed of the baroclinic mode is generally much less than that of the barotropic mode since for average mid-latitude tropospheric conditions $\lambda^2 \simeq 2 \times 10^{-16}$ cm^{-2} which is comparable in magnitude to k^2 for a zonal wavelength of $\sim$4500 km.[1]

As the second special case, we assume that $\beta = 0$. This case corresponds, for example, to a laboratory situation in which the fluid is bounded above and below by rotating horizontal planes so that the gravity and rotation vectors are everywhere parallel. In such a situation

$$c = U_{\mathrm{m}} \pm U_{\mathrm{T}} \left(\frac{k^2 - 2\lambda^2}{k^2 + 2\lambda^2} \right)^{1/2} \tag{10.22}$$

For waves with zonal wave numbers satisfying $k^2 < 2\lambda^2$, (10.20) has an imaginary part. Thus, all waves longer than the critical wavelength $L_{\mathrm{c}} = 2\pi/\sqrt{2\lambda}$ will amplify. Again setting $\lambda \simeq \sqrt{2} \times 10^{-8}$ cm^{-1} we find that $L_{\mathrm{c}} \simeq 3000$ km for typical midlatitude tropospheric conditions. Also from the definition of λ we can write

$$L_{\mathrm{c}} = \frac{\Delta p \, \pi \sqrt{2}}{f_0}$$

Thus, the critical wavelength for baroclinic instability increases with the static stability. The role of static stability in stabilizing the shorter waves can be understood qualitatively as follows: For a sinusoidal perturbation, the relative vorticity, and hence the differential vorticity advection, increases with the square of the wave number. But, as shown in Chapter 7, a secondary vertical circulation is required to maintain hydrostatic temperature changes and geostrophic vorticity changes in the presence of differential vorticity advection. Thus, for a geopotential perturbation of fixed amplitude the relative strength of the accompanying vertical circulation must increase as the wavelength of the disturbance decreases. Since static stability tends to resist vertical displacements, the shortest wavelengths will thus be stabilized.

It is also of interest that with $\beta = 0$ the criterion for instability does not depend on the magnitude of the basic state thermal wind U_{T}. All wavelengths longer than L_{c} are unstable even for very small vertical shear. However, the growth rate of the perturbation does depend on U_{T}. From (10.14) we see that the exponential growth rate is $\alpha = kc_{\mathrm{i}}$, where c_{i} designates the imaginary part of the phase speed. In the present case

---

[1] The presence of the free internal Rossby wave should actually be regarded as a weakness of the two level model. Lindzen et al. (1967) have shown that this mode does not correspond to any free oscillation of the real atmosphere but is, rather, a spurious mode resulting from the use of the upper boundary condition $\omega = 0$ at $p = 0$ which formally turns out to be equivalent to putting a lid at the top of the atmosphere.

$$\alpha = kU_T \left(\frac{2\lambda^2 - k^2}{2\lambda^2 + k^2}\right)^{1/2} \tag{10.23}$$

so that the growth rate increases linearly with the mean thermal wind.

Returning to the general case where all terms are retained in (10.19), the stability criterion is most easily understood by computing the so-called *neutral curve* which connects all values of U_T and k for which $\delta = 0$ so that the flow is *marginally stable*. From (10.19), the condition $\delta = 0$ implies that

$$\frac{\beta^2 \lambda^4}{k^4(k^2 + 2\lambda^2)} = U_T{}^2(2\lambda^2 - k^2) \tag{10.24}$$

This complicated relationship between U_T and k can best be displayed by solving (10.24) for $k^4/2\lambda^4$ yielding,

$$\frac{k^4}{2\lambda^4} = 1 \pm \left[1 - \frac{\beta^2}{4\lambda^2 U_T{}^2}\right]^{1/2}$$

In Fig. 10.1 the nondimensional quantity $k^2/2\lambda^2$, which is a measure of the

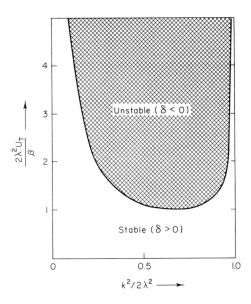

Fig. 10.1 Neutral stability curve for the two-level baroclinic model.

zonal wavelength, is plotted against the nondimensional parameter $2\lambda^2 U_T/\beta$, which is proportional to the thermal wind. As indicated in the figure the neutral curve separates the unstable region of the U_T, k plane from the stable region. It is clear that the inclusion of the β effect serves to stabilize the flow, for now unstable roots exist only for

$$U_{\rm T} > \frac{\beta}{2\lambda^2}$$

In addition the minimum value of $U_{\rm T}$ required for unstable growth depends strongly on k. Thus, the β effect strongly stabilizes the long-wave end of the wave spectrum ($k \to 0$). Again the flow is always stable for waves shorter than the critical wavelength $L_{\rm c} = 2\pi/\sqrt{2\lambda}$.

Differentiating (10.24) with respect to k and setting $dU_{\rm T}/dk = 0$ we find that the minimum value of $U_{\rm T}$ for which unstable waves may exist occurs when $k^2 = \sqrt{2}\lambda^2$. This wave number corresponds to the wave of *maximum instability*. It seems likely that the wave numbers of observed disturbances should be close to the wave number of maximum instability, for if $U_{\rm T}$ were gradually raised from zero the flow would first become unstable for perturbations of wave number $k = 2^{1/4}\lambda$. Those perturbations would then amplify and in the process remove energy from the mean thermal wind, thereby decreasing $U_{\rm T}$ and stabilizing the flow. Under normal conditions of static stability the wavelength of maximum instability is about 4000 km, which is close to the average wavelength for midlatitude synoptic systems. Furthermore, the thermal wind required for marginal stability at this wavelength is only about $U_{\rm T} \simeq 4$ m sec^{-1}, which implies a shear of 8 m sec^{-1} between 250–750 mb. Shears greater than this are certainly common in middle latitudes for the zonally averaged flow. Therefore, the observed behavior of midlatitude synoptic systems is consistent with the hypothesis that such systems can originate from infinitesimal perturbations of a baroclinically unstable basic current. Of course in the real atmosphere many other factors may influence the development of synoptic systems, for example, instabilities due to lateral shear in the jet stream, nonlinear interactions of finite amplitude perturbations, and the release of latent heat. However, the evidence from observational studies, laboratory simulations, and numerical models all suggests that baroclinic instability is the primary mechanism for synoptic scale wave development in middle latitudes.

10.3   The Energetics of Baroclinic Waves

We have seen in the previous section that under suitable conditions a basic state current which contains vertical shear will be unstable to small perturbations. Such perturbations can then amplify exponentially by drawing potential and/or kinetic energy from the mean flow. In this section we analyze the energetics of linearized baroclinic disturbances and show that these perturbations grow by converting potential energy of the mean flow.

10.3.1 AVAILABLE POTENTIAL ENERGY

Before discussing the energetics of baroclinic waves, it is necessary to consider the energy of the atmosphere from a more general point of view. For all practical purposes, the total energy of the atmosphere is the sum of the internal energy, gravitational potential energy, and kinetic energy. However, it is not necessary to consider separately the variations of internal and gravitational potential energy because in a hydrostatic atmosphere these two forms of energy are proportional and may be combined into a single term called the *total potential energy*. The proportionality of internal and gravitational potential energy may be shown simply by considering these forms of energy for a column of air of unit horizontal cross section which extends from the surface to the top of the atmosphere.

If we let dE_1 be the internal energy in a vertical section of the column of height dz, then from the definition of internal energy

$$dE_1 = \rho c_v T \, dz$$

so that the internal energy for the entire column is

$$E_1 = c_v \int_0^\infty \rho T \, dz \tag{10.25}$$

On the other hand, the gravitational potential energy for a slab of thickness dz at a height z is just

$$dE_p = \rho g z \, dz$$

so that the gravitational potential energy in the entire column is

$$E_p = \int_0^\infty \rho g z \, dz = - \int_{p_0}^0 z \, dp \tag{10.26}$$

where we have substituted from the hydrostatic equation to obtain the last integral in (10.26). Integrating (10.26) by parts and using the ideal gas law we obtain

$$E_p = \int_0^\infty p \, dz = R \int_0^\infty \rho T \, dz \tag{10.27}$$

Comparing (10.25) and (10.27) we see that

$$c_v E_p = R E_1 \tag{10.28}$$

Thus, the total potential energy may be expressed as

$$E_p + E_1 = \frac{c_p}{c_v} E_1 = \frac{c_p}{R} E_p \tag{10.29}$$

Therefore, in a hydrostatic atmosphere the total potential energy can be obtained by computing either E_1 or E_p alone.

The total potential energy is not a very suitable measure of energy in the atmosphere because only a very small fraction of the total potential energy is available for conversion to kinetic energy in storms. To understand qualitatively why most of the total potential energy is unavailable we consider a simple model atmosphere which initially consists of two dry air masses separated by a vertical partition as shown in Fig. 10.2. The two air masses

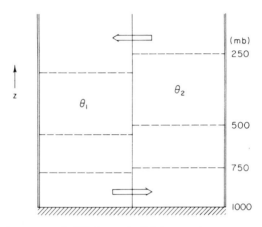

Fig. 10.2 Two air masses of differing potential temperature separated by a vertical partition. Dashed lines indicate isobaric surfaces. Arrows show direction of motion when partition is removed.

are at uniform potential temperatures θ_1 and θ_2, respectively, with $\theta_1 < \theta_2$. The ground level pressure on each side of the partition is taken to be 1000 mb. We now wish to compute the maximum kinetic energy which can be realized by an *adiabatic* rearrangement of mass within the same volume when the partition is removed. Now for an adiabatic process, total energy is conserved:

$$E_k + E_p + E_1 = \text{const}$$

where E_k denotes the kinetic energy. If the air masses are initially at rest $E_k = 0$. Thus, if we let primed quantities denote the final state

$$K_k' + E_p' + E_1' = E_p + E_1$$

so that with the aid of (10.29) we find that the kinetic energy realized by removal of the partition is

$$E_k' = \frac{c_p}{c_v}(E_1 - E_1') \tag{10.30}$$

Since θ is conserved for an adiabatic process, no mixing is allowed. It is clear that E_1' will be a minimum (designated by E_1'') when the masses are rearranged so that the air at θ_1 lies entirely beneath the θ_2 air with the 500-mb surface as the horizontal boundary between the two masses. In that case the total potential energy

$$\frac{c_p}{c_v} E_1''$$

is not available for conversion to kinetic energy, because no adiabatic process can further reduce E_1''.

The *available potential energy* (often abbreviated as APE) can now be defined as the difference between the total potential energy of a closed system and the minimum total potential energy which could result from an adiabatic redistribution of mass. Thus, for the idealized model given above

$$P \equiv \frac{c_p}{c_v}(E_1 - E_1'') \qquad (10.31)$$

which is equivalent to the maximum kinetic energy which can be realized by an adiabatic process.

Lorenz (1960) has shown that for the earth's atmosphere, the available potential energy is given approximately by the volume integral over the entire atmosphere of the variance of potential temperature on isobaric surfaces. Thus, letting $\bar{\theta}$ designate the average potential temperature for a given pressure surface, and θ' the local deviation from the average, the average available potential energy per unit volume satisfies the proportionality

$$\bar{P} \propto \frac{1}{V} \int \frac{\overline{\theta'^2}}{\bar{\theta}^2}\, dV$$

where V designates the total volume. For the quasi-geostrophic model, this proportionality is an *exact* measure of the available potential energy, as we will show in the following subsection.

Observations indicate that for the atmosphere as a whole

$$\frac{\bar{P}}{(c_p/c_v)\bar{E}_1} \sim \frac{1}{200}$$

$$\frac{\bar{K}}{\bar{P}} \sim \frac{1}{10}$$

Thus only about 0.5 percent of the total potential energy of the atmosphere is available, and of the portion available only about 10 percent is actually converted to kinetic energy. From this point of view the atmosphere is a rather inefficient heat engine.

10.3.2 ENERGY EQUATIONS FOR THE TWO-LEVEL LINEARIZED QUASI-GEOSTROPHIC MODEL

In our two-level model the perturbation temperature field is proportional to $\psi_1' - \psi_3'$, the 250–750-mb thickness. Thus, in view of the discussion in the previous section, we anticipate that the available potential energy in this case is proportional to $(\psi_1' - \psi_3')^2$. To show that this in fact must be the case, we derive the energy equations for the system in the following manner: We first multiply (10.11) by $-\psi_1'$, (10.12) by $-\psi_3'$, and (10.13) by $\psi_1' - \psi_3'$. We then integrate the resulting equations over one wavelength of the perturbation in the zonal direction. The resulting zonally averaged terms will be denoted by the angle bracket notation

$$\langle (\ ) \rangle = \frac{1}{L} \int_0^L (\ ) \, dx$$

where L is the wavelength of the perturbation. Thus, for the first term in (10.11) we have after multiplying by $-\psi_1'$:

$$-\left\langle \psi_1' \frac{\partial}{\partial t} \left(\frac{\partial^2 \psi_1'}{\partial x^2} \right) \right\rangle = -\left\langle \psi_1' \frac{\partial^2}{\partial x^2} \left(\frac{\partial \psi_1'}{\partial t} \right) \right\rangle$$

$$= \underbrace{-\left\langle \frac{\partial}{\partial x} \left[\psi_1' \frac{\partial}{\partial x} \left(\frac{\partial \psi_1'}{\partial t} \right) \right] \right\rangle}_{A} + \underbrace{\left\langle \frac{\partial \psi_1'}{\partial x} \frac{\partial}{\partial t} \left(\frac{\partial \psi_1'}{\partial x} \right) \right\rangle}_{B}$$

Term A vanishes because it is the integral of a perfect differential in x. Term B can be rewritten as

$$\left\langle \frac{1}{2} \frac{\partial}{\partial t} \left(\frac{\partial \psi_1'}{\partial x} \right)^2 \right\rangle$$

which is just the rate of change of the perturbation kinetic energy averaged over a wavelength. Similarly, $-\psi_1'$ times the advection term on the right in (10.11) can be written after integration in x as

$$-U_1 \left\langle \psi_1' \frac{\partial^2}{\partial x^2} \left(\frac{\partial \psi_1'}{\partial x} \right) \right\rangle = -U_1 \left\langle \frac{\partial}{\partial x} \left[\psi_1' \frac{\partial}{\partial x} \left(\frac{\partial \psi_1'}{\partial x} \right) \right] \right\rangle + U_1 \left\langle \frac{\partial \psi_1'}{\partial x} \frac{\partial^2 \psi_1'}{\partial x^2} \right\rangle$$

$$= +\frac{U_1}{2} \left\langle \frac{\partial}{\partial x} \left[\frac{\partial \psi_1'}{\partial x} \right]^2 \right\rangle = 0$$

Thus, the advection of kinetic energy vanishes when integrated over a wavelength. Evaluating the various terms in (10.12) and (10.13) in the same manner after multiplying through by $-\psi_3'$ and $(\psi_1' - \psi_3')$, respectively, we obtain the following set of perturbation energy equations:

$$\frac{1}{2}\left\langle \frac{d}{dt}\left(\frac{\partial \psi_1'}{\partial x}\right)^2 \right\rangle = -\frac{f_0}{\Delta p}\langle \omega_2'\psi_1' \rangle \tag{10.32}$$

$$\frac{1}{2}\left\langle \frac{d}{dt}\left(\frac{\partial \psi_3'}{\partial x}\right)^2 \right\rangle = \frac{f_0}{\Delta p}\langle \omega_2'\psi_3' \rangle \tag{10.33}$$

$$\frac{1}{2}\left\langle \frac{d}{dt}(\psi_1' - \psi_3')^2 \right\rangle = U_T\left\langle (\psi_1' - \psi_3')\frac{\partial}{\partial x}(\psi_1' + \psi_3') \right\rangle$$

$$+ \frac{\sigma\,\Delta p}{f_0}\langle \omega_2'(\psi_1' - \psi_3') \rangle \tag{10.34}$$

where as before $U_T \equiv (U_1 - U_3)/2$.

Defining the perturbation kinetic energy to be the sum of the kinetic energies of the 250 and 750 mb levels:

$$K' \equiv \frac{1}{2}\left\langle \left(\frac{\partial \psi_1'}{\partial x}\right)^2 \right\rangle + \frac{1}{2}\left\langle \left(\frac{\partial \psi_3'}{\partial x}\right)^2 \right\rangle$$

we find by adding (10.32) and (10.33) that

$$\frac{dK'}{dt} = -\frac{f_0}{\Delta p}\langle \omega_2'(\psi_1' - \psi_3') \rangle \tag{10.35}$$

Thus, the rate of change of perturbation kinetic energy is proportional to the correlation between the perturbation thickness and vertical motion. If we now define the perturbation available potential energy as

$$P' = \frac{\lambda^2\langle (\psi_1' - \psi_3')^2 \rangle}{2}$$

we obtain from (10.34)

$$\frac{dP'}{dt} = \lambda^2 U_T\left\langle (\psi_1' - \psi_3')\frac{\partial}{\partial x}(\psi_1' + \psi_3') \right\rangle + \frac{f_0}{\Delta p}\langle \omega_2'(\psi_1' - \psi_3') \rangle \tag{10.36}$$

The last term in (10.36) is just equal and opposite to the kinetic-energy source term in (10.35). This term clearly must represent a conversion between potential and kinetic energy. If on the average the vertical motion is positive ($\omega_2' < 0$) where the thickness is greater than average ($\psi_1' - \psi_3' > 0$) and vertical motion is negative where thickness is less than average, we have

$$\langle \omega_2'(\psi_1' - \psi_3') \rangle < 0 \tag{10.37}$$

perturbation potential energy is then being converted to kinetic energy. Physically, this correlation represents an overturning in which warm air is rising and cold air sinking, a situation which clearly tends to lower the center of mass and hence the potential energy of the perturbation. However, the available potential energy and kinetic energy of a disturbance can still grow simultaneously, provided that the potential-energy generation due to the first term in (10.36) exceeds the rate of potential energy conversion to kinetic energy.

The potential-energy generation term in (10.36) depends on the correlation between the perturbation thickness $(\psi_1' - \psi_3')$ and meridional velocity $\partial(\psi_1' + \psi_3')/\partial x$ at 500 mb. In order to understand the role of this term, it is helpful to consider a particular sinusoidal wave disturbance. Suppose that the barotropic and baroclinic parts of the disturbance can be written respectively as

$$\psi_1' + \psi_3' = A_M \cos k(x - ct)$$
$$\psi_1' - \psi_3' = A_T \cos k(x + x_0 - ct)$$

(10.38)

where x_0 designates the phase difference. Since $\psi_1' + \psi_3'$ is proportional to the 500-mb geopotential, and $\psi_1' - \psi_3'$ is proportional to the 500-mb temperature, the phase angle kx_0 gives the phase difference between the geopotential and temperature fields at 500 mb. Furthermore, A_M and A_T are measures of the amplitudes of the 500-mb-disturbance geopotential and temperature fields, respectively. Using the expressions in (10.38) we obtain

$$\left\langle (\psi_1' - \psi_3') \frac{\partial}{\partial x} (\psi_1' + \psi_3') \right\rangle$$

$$= \frac{k}{L} \int_0^L A_T A_M \cos k(x + x_0 - ct) \sin k(x - ct)\, dx$$

$$= \frac{k A_T A_M \sin kx_0}{L} \int_0^L [\sin k(x - ct)]^2\, dx$$

$$= \frac{A_T A_M k \sin kx_0}{L}$$

(10.39)

From (10.36) we see that for the usual midlatitude case of a westerly thermal wind $(U_T > 0)$ the correlation in (10.39) must be positive if the perturbation potential energy is to increase. Thus, x_0 must satisfy

$$0 < kx_0 < \pi$$

Furthermore, the correlation will be a positive maximum for $kx_0 = \pi/2$, that is, when the *temperature wave lags the geopotential wave by* 90° at 500 mb. This case is shown schematically in Fig. 10.3. Clearly, when the temperature wave lags the geopotential by one-quarter cycle, the northward advection of warm air by the geostrophic wind east of the 500-mb trough, and the south-

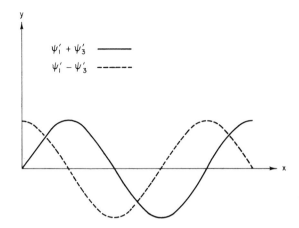

Fig. 10.3 Schematic illustration of the phase relationship between the temperature and geopotential disturbances at 500 mb for maximum perturbation potential energy generation.

ward advection of cold air west of the 500-mb trough are both maximized. Since the disturbance amplitude at 250 mb is generally greater than the amplitude at 750 mb, the $\psi_1' - \psi_3'$ field tends to be nearly in phase with ψ_1'. As a result, the cold advection in the lower troposphere is strong below the 250-mb trough and the warm advection in the lower troposphere is strong below the 250-mb ridge. In that case, as discussed previously in Section 7.3.1, the upper-level disturbance will intensify. It should also be noted here that if the temperature wave lags the geopotential wave, the trough and ridge axes will tilt westward with height which, as mentioned in Section 7.1, is observed to be the case for amplifying midlatitude synoptic systems.

Referring again to Fig. 10.3 and recalling the vertical motion pattern implied by the omega equation (8.34), we see that the signs of the two terms on the right in (10.36) cannot be the same. For in the westward tilting perturbation of Fig. 10.3, the vertical motion must be downward in the cold air behind the trough at 500 mb and upward in the warm air ahead of the trough in order that the divergent motions can produce the necessary geostrophic vorticity change at 250 mb. Hence, the correlation between temperature and vertical velocity must be positive in this situation. Thus, for quasi-geostrophic perturbations, a westward tilt of the perturbation with height implies *both* that the horizontal temperature advection will increase the available potential energy of the perturbation and that the vertical circulation will convert perturbation available potential energy to perturbation kinetic energy. Conversely, an eastward tilt of the system with height would change the direction of both terms on the right in (10.36).

Although the signs of the potential-energy generation term and the potential-energy conversion term in (10.36) are always opposite for a developing baroclinic wave, it is only the potential-energy generation rate which determines the growth of the total energy, $P' + K'$, of the disturbance. This may be proved by adding (10.35) and (10.36) to obtain

$$\frac{d}{dt}(P' + K') = \lambda^2 U_\mathrm{T} \left\langle (\psi_1' - \psi_3') \frac{\partial}{\partial x} (\psi_1' + \psi_3') \right\rangle$$

Thus, provided the correlation between the meridional velocity and temperature is positive and $U_\mathrm{T} > 0$, the total energy of the perturbation will increase. Note that the vertical circulation merely converts disturbance energy between the available potential and kinetic forms without affecting the total energy of the perturbation.

The rate of increase of the total energy of the perturbation depends on the magnitude of U_T, that is, on the zonally averaged meridional temperature gradient. Since the generation of perturbation energy requires systematic northward transport of warm air and southward transport of cold air, it is clear that these disturbances will tend to reduce the meridional temperature gradient and hence the available potential energy of the mean flow. This latter process cannot be mathematically described in terms of the linearized equations. However, from Fig. 10.4 we can see qualitatively that for pertur-

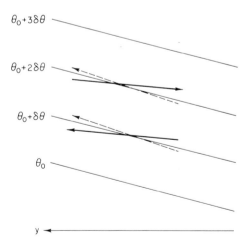

Fig. 10.4 Slopes of parcel trajectories relative to the zonal mean potential temperature surfaces for baroclinically unstable disturbances (solid arrows) and for a baroclinically stable disturbance (dashed arrows).

bations to extract potential energy from the mean flow the perturbation parcel trajectories in the meridional plane must have slopes less than the slopes of the potential temperature surfaces. Since we have previously seen that north-ward-moving air must rise and southward-moving air must sink, it is clear that the rate of energy generation can be greater for an atmosphere in which the meridional slope of the potential temperature surfaces is large. We can also see more clearly why there is a short-wave cut off from baroclinic insta-bility. As previously mentioned, the intensity of the vertical circulation must increase as the wavelength decreases. Thus, the slopes of the parcel trajectories must increase with decreasing wavelength, and for some critical wavelength the trajectory slopes will become greater than the slopes of the potential temperature surfaces.

The energy flow for quasi-geostrophic perturbations is summarized in Fig. 10.5 by means of a block diagram. In this type of energy diagram each

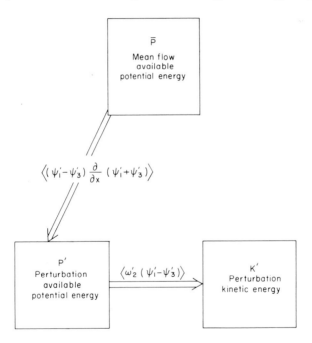

Fig. 10.5 Energy flow in an amplifying baroclinic wave.

block represents a reservoir of a particular type of energy and the arrows designate the direction of energy flow. The complete energy cycle cannot be derived in terms of linear perturbation theory, but will be discussed quali-tatively in Chapter 11.

10.4   Fronts and Frontogenesis

In the above discussion of baroclinic instability, we assumed that the mean thermal wind U_T was a constant independent of the y coordinate. This assumption was necessary to obtain a mathematically simple model which could still illustrate the basic instability mechanism. However, it was pointed out in Section 7.1 that baroclinicity is not uniformly distributed, but rather the horizontal temperature gradients tend to be concentrated in narrow frontal zones associated with the tropospheric jet streams. Thus in the real atmosphere the thermal wind is strongly latitudinally dependent with a maximum at the latitude of the jet core. This lateral dependence of the thermal wind must be included in any complete theory of baroclinic wave development. However, the fact that the simple theory of the previous section was able to qualitatively reproduce the essential features of developing baroclinic waves indicates that the fundamental process in the generation of midlatitude storms is baroclinic instability due to vertical shear, not some process associated with the lateral shear of the zonal wind across the jet stream.

We have shown in the previous section that the energetics of baroclinic waves require that the waves remove available potential energy from the mean flow. Thus, baroclinic wave development will tend to weaken the mean meridional temperature gradient (that is, reduce the mean thermal wind). The mean pole-to-equator temperature gradient is of course continually restored by differential solar heating. However, differential heating cannot account for the tendency of the meridional temperature gradient to be concentrated along the polar front. Clearly, some dynamic process is required which can continuously reestablish the strong gradients characteristic of the frontal zone. Such a process is called *frontogenetic*.

10.4.1   THE KINEMATICS OF FRONTOGENESIS

Although a completely satisfactory dynamic theory of frontogenesis has yet to be developed, some qualitative notion of the conditions favorable for frontogenesis can be obtained by a strictly *kinematic* analysis, that is, a description of the geometry of the flow without reference to the underlying physical forces.

It can easily be seen from inspection that a velocity field such as that shown in Fig. 10.6 will tend to advect the temperature field so that the isotherms are concentrated along the so-called axis of dilation (the x axis in Fig. 10.6), provided that the initial temperature field has a finite gradient along the axis of contraction (the y axis in Fig. 10.6). The velocity field shown in Fig. 10.6 is a pure *deformation* field which has a streamfunction given by $\psi = -Kxy$. It is easily shown that a pure deformation field has zero vorticity

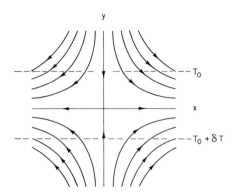

Fig. 10.6 The streamfunction $\psi = -Kxy$ for a pure deformation field.

and zero divergence. A parcel advected by a deformation field will merely have its shape changed in time, without any change in horizontal area. For example, a square parcel with sides parallel to the x and y axes would be deformed into a rectangle, as should be obvious from Fig. 10.6.

A pure deformation field does not resemble any typical synoptic scale atmospheric field. However, if a large component of mean zonal flow is added to the deformation field, a *confluent* flow will result in the region near the origin. Such confluent regions are always present in the tropospheric jet stream due to the influence of planetary scale waves on the position and intensity of the jet. In fact, even a monthly mean chart, such as in Fig. 7.3, reveals two regions of large-scale confluence at 500 mb. Observationally, confluence of this sort is known to be associated with frontogenesis.

In a homogeneous incompressible fluid a confluent flow such as shown in Fig. 10.7a would be nondivergent. However, in the atmosphere the increase in concentration of the isotherms along the jet requires an increase in the thermal wind in the jet. As a result, the cyclonic vorticity north of the jet core and anticyclonic vorticity south of the jet core must both increase. In order to produce these vorticity changes a secondary circulation such as shown in Fig. 10.7b is required. Upper-level convergence north of the jet is required to produce sufficient cyclonic vorticity to keep the vorticity geostrophic. Conversely, upper-level divergence is required south of the jet to produce anticyclonic vorticity. The vertical circulation required by mass continuity is also indicated in Fig. 10.7b. Again, just as was the case for baroclinic waves, the vertical circulation produces adiabatic temperature changes which tend to oppose the horizontal temperature advection. Thus, the vertical circulation reduces the rate of upper-level frontogenesis (for quasi-geostrophic motions).

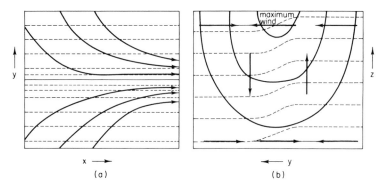

Fig. 10.7 (a) Horizontal streamlines and isotherms in a frontogenetic confluence. (b) Vertical section across the confluence, showing isotachs (solid), isotherms (dashed), and vertical and transverse motions (arrows). (After Sawyer, 1956.)

10.4.2 QUASI-GEOSTROPHIC FRONTOGENESIS

We can demonstrate that a confluent flow such as shown in Fig. 10.7a will indeed tend to concentrate a small preexisting temperature gradient. For this purpose a simple theoretical analysis using our two-level quasi-geostrophic model will suffice.[2]

The quasi-geostrophic model is, of course, only valid when the horizontal scale of the temperature variation is large enough so that the scaling assumptions are not violated. Thus, this model cannot describe a fully developed front, for in that case the horizontal scale is $\lesssim 100$ km and the twisting and vertical advection terms can no longer be neglected in the vorticity equation. However, the quasi-geostrophic model can be used to simulate the initial stages of frontogenesis before the lateral scale has become subsynoptic.

We assume in this model that the initial streamfunction field contains a deformation component which is height independent plus a perturbation which is periodic in y. Thus, we let

$$\psi_1 = -Kxy + \psi_1'(y, t)$$
$$\psi_3 = -Kxy + \psi_3'(y, t)$$

(10.40)

where K is a constant with dimensions of inverse time which describes the strength of the deformation field. We let $\psi_m = \psi_1' + \psi_3'$ and $\psi_T = \psi_1' - \psi_3'$ designate the barotropic and baroclinic parts of the perturbation field, respectively. We then obtain from (10.40)

[2] A similar analysis for the continuous atmosphere was given by Williams and Plotkin (1968).

$$\psi_1 + \psi_3 = -2Kxy + \psi_m(y, t)$$
$$\psi_1 - \psi_3 = \psi_T(y, t) \tag{10.41}$$

Substituting (10.41) into the two-level model equations (8.31) and (8.32) we obtain

$$\left(\frac{\partial}{\partial t} - Ky\frac{\partial}{\partial y}\right)\left(\frac{\partial^2 \psi_m}{\partial y^2}\right) = 0 \tag{10.42}$$

$$\left(\frac{\partial}{\partial t} - Ky\frac{\partial}{\partial y}\right)q_T = 0 \tag{10.43}$$

where

$$q_T \equiv \left(\frac{\partial^2}{\partial y^2} - 2\lambda^2\right)\psi_T$$

is the potential vorticity of the thermal wind. Equations (10.42) and (10.43) state that the vertically averaged vorticity

$$\frac{\partial^2 \psi_m}{\partial y^2}$$

and the potential vorticity of the thermal wind q_T are both conserved following the motion of the deformation field. Furthermore, they are both advected at the speed $-Ky$ in the y direction. Thus, it is obvious that an initial gradient in q_T will be concentrated toward $y = 0$.

Following the motion of an individual parcel, we have

$$\frac{dy}{dt} = -Ky \qquad \text{or} \qquad \frac{dy}{y} = -K\,dt$$

which may be integrated to give the position of a parcel as

$$y = y_0 e^{-Kt}$$

where y_0 is the initial position of the parcel. Thus, we can write the distributions of q_T and ψ_m at any time in terms of the initial distributions:

$$q_T(y, t) = q_T(ye^{Kt}, 0)$$

$$\frac{\partial^2 \psi_m}{\partial y^2}(y, t) = \frac{\partial^2 \psi_m}{\partial y^2}(ye^{Kt}, 0) \tag{10.44}$$

We now assume as the initial temperature distribution a sinusoidal field

$$\psi_T = -A_T \sin my$$

In that case

$$q_T(y, 0) = A_T(m^2 + 2\lambda^2)\sin my \tag{10.45}$$

Hence $\psi_T(y, t)$ must satisfy

$$\left(\frac{\partial^2}{\partial y^2} - 2\lambda^2\right)\psi_T = A_T(m^2 + 2\lambda^2)\sin(me^{Kt}y) \tag{10.46}$$

The solution of (10.46) which satisfies the initial condition is

$$\psi_T = -\frac{A_T(m^2 + 2\lambda^2)}{(m^2 e^{2Kt} + 2\lambda^2)}\sin(me^{Kt} y) \tag{10.47}$$

From (10.47) we see that the lateral scale of the temperature perturbation decreases exponentially in time (that is, the wave number increases exponentially as e^{Kt}). At the same time, however, the maximum amplitude of ψ_T decreases because of the e^{2Kt} factor in the denominator. This unrealistic feature arises from our choice of an initially sinusoidal temperature field plus a constant deformation field. Of primary interest in the present discussion is the temperature gradient (or perturbation thermal wind) given by

$$-\frac{\partial \psi_T}{\partial y} = \frac{A_T me^{Kt}(m^2 + 2\lambda^2)}{(m^2 e^{2Kt} + 2\lambda^2)}\cos(me^{Kt}y) \tag{10.48}$$

As a particular example, we choose constants as follows: We let $m = 0.5 \times 10^{-8}$ cm^{-1} corresponding to an initial distance from the origin to the temperature maximum or minimum of $y \simeq 3000$ km. We let $\lambda^2 = 4 \times 10^{-16}$ cm^{-2}, corresponding to typical tropospheric stability conditions. We set $A_T = 10^{11}$ cm^2 sec^{-1}, which gives an initial thermal wind of 5 m sec^{-1} at $y = 0$. Finally, we let $K = 10^{-5}$ sec^{-1}, corresponding to a deformation velocity along the y axis of -10 m sec^{-1} at $y = \pm 1000$ km. With this choice of K the lateral scale decreases by e^{-1} in about one day, which is a reasonable time scale for quasi-geostrophic frontogenesis. The amplitude of the thermal wind at $y = 0$ as a function of time given by (10.48) is plotted for this example in Fig. 10.8. In the first day the deformation field assumed here has more than doubled the thermal wind (and temperature gradient) at $y = 0$. In this oversimplified model

$$\frac{\partial \psi_T}{\partial y} \rightarrow 0 \qquad \text{as} \quad t \rightarrow \infty$$

However, the assumptions of quasi-geostrophic theory are only valid for $Kt \lesssim 1$, because for larger t the lateral scale is reduced to the point where nongeostrophic processes become important.

From this simple analysis, we can conclude that a moderate deformation field can concentrate a small preexisting temperature gradient on a time scale consistent with observed large-scale frontogenesis. However, this analysis cannot simulate the formation of a true front because the scaling assumptions

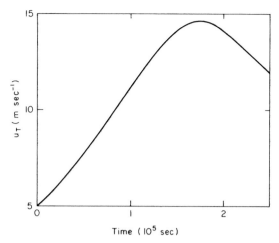

Fig. 10.8 Amplitude of the thermal wind at $y = 0$ as a function of time in the two-level quasi-geostrophic frontogenesis model (see text).

in quasi-geostrophic theory do not apply to a true front. Furthermore, this model applies only to deep frontogenesis in the free atmosphere. In addition, there are intense surface fronts which are often found in association with synoptic wave disturbances. Such processes as surface friction and vertical motion associated with precipitation are likely to be important for the formation and maintenance of these small-scale transient features. However, at present no satisfactory theory exists which incorporates all the physical mechanisms that may be important for frontogenesis.

10.4.3 STEADY-STATE FRONTAL ZONES

One aspect of fronts which could not be incorporated in our two-level model is the characteristic observed slope of the zone of maximum temperature gradient with respect to height (see Fig. 7.4). That a stationary frontal zone must slope with height can be seen most easily by examining the limiting case of a frontal zone embedded in a purely geostrophic wind field which blows parallel to the isotherms. We assume that the entire temperature gradient is confined to a narrow zone of width δy with a horizontal temperature change of δT across the zone, and that outside this zone the temperature is constant in isobaric surfaces. In that case, the thermal wind equation (3.29) states approximately that

$$\frac{\delta U_g}{\delta \ln p} \simeq \frac{R \, \delta T}{f \, \delta y}$$

Thus, if the frontal zone were vertically oriented as shown in Fig. 10.9a there

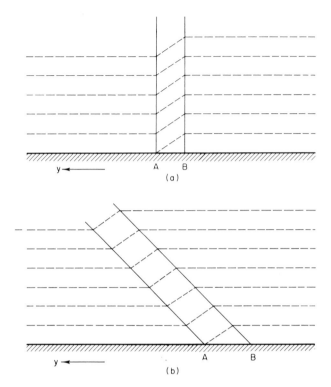

Fig. 10.9 Stationary frontal zones: (a) unstable configuration; (b) stable configuration. Dashed lines indicate isotherms.

would be discontinuities in U_g along the lines A and B since only in the zone between A and B would U_g have vertical shear. The resultant infinite lateral shear along A and B would be a highly unstable configuration and would soon be destroyed by the rapid growth of unstable perturbations along the shear line. Hence, a vertically oriented frontal surface is not a stable configuration. On the other hand, if the frontal zone slopes as shown in Fig. 10.9b, the thermal wind can provide a smooth transition between the constant geostrophic wind U_g in the cold section and the constant geostrophic wind $U_g + \delta U_g$ in the warm section. The frontal slope can then be estimated directly from the thermal wind relationship as

$$\frac{\delta \ln p}{\delta y} \simeq \frac{+f \delta U_g}{R \delta T}$$

or in terms of height coordinates

$$\frac{\delta z}{\delta y} \simeq \frac{-f H \delta U_g}{R \delta T} \tag{10.49}$$

where H is the mean scale height, $H = R\overline{T}/g$. Equation (10.49), although it results from a highly simplified model, does produce a reasonable estimate for the observed slopes of typical midlatitude fronts.

Problems

1. Show using Eq. (10.23) that the maximum growth rate for baroclinic instability when $\beta = 0$ occurs for

$$k^2 = 2\lambda^2(\sqrt{2} - 1)$$

How long does it take the most rapidly growing wave to amplify by a factor of e^1 if $\lambda = \sqrt{2} \times 10^{-8}$ cm^{-1} and $U_T = 20$ m sec^{-1}?

2. Obtain a formula for the Rossby wave phase speed in a homogeneous barotropic fluid confined between rigid horizontal lids when friction is included in the form $\mathbf{Fr} = -\mu \mathbf{V}$ where μ is a constant drag coefficient, and $\mathbf{Fr}$ designates the horizontal friction force. Describe qualitatively the behavior of this sort of wave.

3. Show by substituting the baroclinic Rossby wave speed (10.21) into the perturbation equations (10.15)–(10.17) that the amplitudes of the co-efficients in the assumed solution (10.14) must be related as follows:

$$A = -B, \qquad C = \frac{2ikf_0\beta}{\sigma \Delta p\,(k^2 + 2\lambda^2)}\,A$$

4. Suppose that a baroclinic fluid is confined between two rigid horizontal lids in a rotating tank in which $\beta = 0$ but frictional drag must be included. If the frictional force has the same form as given in Problem 2 (that is, friction is everywhere linearly proportional to the velocity, but not dependent on shear) show that the two-level model perturbation vorticity equations in cartesian coordinates can be written as

$$\left(\frac{\partial}{\partial t} + U_1 \frac{\partial}{\partial x} + \mu\right) \frac{\partial^2 \psi_1'}{\partial x^2} - \frac{f}{\Delta p}\,\omega_2 = 0$$

$$\left(\frac{\partial}{\partial t} + U_3 \frac{\partial}{\partial x} + \mu\right) \frac{\partial^2 \psi_3'}{\partial x^2} + \frac{f}{\Delta p}\,\omega_2 = 0$$

where perturbations are assumed in the form given in (10.10). Assuming solutions of the form (10.14) show that the phase speed satisfies a relationship similar to (10.19) with β replaced everywhere by $i\mu k$, and that as a result the condition for baroclinic instability becomes

$$U_T \geq \frac{\mu}{(2\lambda^2 - k^2)^{1/2}}$$

Plot the marginal stability curve using the ordinate $\lambda U_T/\mu$ and the abscissa $k^2/2\lambda^2$. Explain this result physically.

5. Verify that the expression for $q_T(y, t)$ in (10.44) is a solution for (10.43) by direct substitution. *Hint:* Transform variables by letting $\eta = e^t y$ and differentiate $q_T(\eta, 0)$ using the chain rule.

6. Compute the slope of a stationary geostrophic frontal zone at 43°N if the temperature and geostrophic wind change across the front by 10°C and 30 m sec^{-1}, respectively. Assume a mean scale height of 8 km.

7. Verify that the phase speeds given by (10.19) are roots of the quadratic (10.18).

Suggested References

Palmén and Newton, *Atmospheric Circulation Systems*, describes the observational aspects of baroclinic wave development and frontogenesis.

Phillips (1963) is a review paper on many aspects of geostrophic motions including a brief account of the principal results of baroclinic instability theory. This paper also contains an extensive bibliography of original literature on the subject.

Charney (1947) is the classical paper on baroclinic instability. The mathematical level is advanced, but Charney's paper contains an excellent qualitative summary of the main results which is very readable.

Chapter

11 | The General Circulation

The general circulation of the atmosphere is usually considered to consist of those components of the motion which remain when the flow is averaged in time or in space, or both. Most general circulation research, both observational and theoretical, has emphasized the zonally averaged circulation, that is, the latitude–height distribution of various flow parameters averaged around latitude circles. This approach is especially relevant to theoretical studies where, as we have previously seen, it is very useful to divide the flow into zonal mean and perturbation components. However, zonal averaging excludes the possibility of studying the so-called monsoonal circulations which are longitudinally dependent *time averaged* circulations forced by topography and continent–ocean heating contrasts. Although the monsoonal circulations strongly influence the earth's climate, we will omit discussion of them here since no simple theoretical treatment exists.

In the present chapter we will limit the discussion to the zonally averaged circulation only. This approach allows us to focus on those features which are not dependent on orography, and should thus be common to all thermally driven rotating atmospheres. The object of the present chapter is then to show that the observed structure of the zonally averaged general circulation is consistent with the constraints imposed by the dynamical principles discussed in the previous chapters. And, in particular, we will show that in middle latitudes the general circulation is qualitatively in accord with the predictions of quasi-geostrophic theory.

11.1 The Nature of the Problem

Theoretical speculation on the nature of the general circulation has quite a long history. Perhaps the most important early work on the subject was that of eighteenth-century Englishman, George Hadley. Hadley, in seeking a cause for the trade wind circulation, realized that this circulation must be a form of thermal convection driven by the difference in solar heating between the equatorial and polar regions. He visualized the general circulation as the result of a zonally symmetric overturning in which the heated equatorial air rises and moves poleward where it cools, sinks, and moves equatorward again. This type of circulation is now called a *Hadley circulation* or Hadley cell.

Although the Hadley circulation is a mathematically possible circulation in the sense that it does not violate any of the laws of physics, it is not observed to be the primary mode of circulation for the earth's atmosphere. Evidence from a number of types of studies indicates that under the conditions existing in the earth's atmosphere, a symmetric *Hadley* circulation would be baroclinically unstable. The observed general circulation may then be thought of qualitatively as developing in the following way:

In the mean the net solar energy absorbed by the atmosphere and the earth must equal the infrared energy radiated back to space by the planet. However, solar heating is strongly dependent on latitude, with a maximum at the equator and a minimum at the poles. The outgoing infrared radiation, on the other hand, is only weakly latitude dependent. Thus, there is a radiation excess in the equatorial region and a deficit in the polar region. This differential heating creates a pole-to-equator temperature gradient, and hence produces a growing store of zonal mean available potential energy. However, at some point the zonal thermal wind (which must develop if the motion is to be geostrophically balanced in the presence of the pole-to-equator temperature gradient) becomes baroclinically unstable. As shown in the previous chapter, the resulting baroclinic waves transport heat northward. These waves will intensify until their heat transport is sufficient to balance the radiation deficit in the polar regions so that the pole-to-equator temperature gradient ceases to grow. At the same time, these perturbations convert potential energy into kinetic energy, thereby maintaining the kinetic energy of the atmosphere against the effects of frictional dissipation.

From a thermodynamic point of view then, the atmosphere may be regarded as a "heat engine" which absorbs net heat at a high-temperature "reservoir" in the tropics and gives up heat from a low-temperature reservoir in the polar regions. In this manner radiation generates available potential energy which is in turn partially converted to kinetic energy which does work against friction. However, only a small fraction of the solar energy input actually gets converted to kinetic energy. Thus, from an engineer's viewpoint the atmosphere is a rather inefficient heat engine. However, if due account is taken of the many constraints operating on atmospheric motions, it appears that the atmosphere may in fact generate kinetic energy about as efficiently as dynamically possible.

The above qualitative discussion suggests that the gross features of the general circulation can be understood on the basis of quasi-geostrophic theory since, as we have previously seen, baroclinic instability processes are contained within the quasi-geostrophic framework. To keep the discussion as simple as possible, we will in this chapter concentrate on those aspects of the general circulation which are contained in the quasi-geostrophic model. Thus, we will focus primarily on the circulation of a dry atmosphere outside the tropics. The general circulation within the tropics cannot be understood without considering latent heat release in convective clouds. This subject will be postponed until Chapter 12.

It should be recognized that a quais-geostrophic model cannot provide a complete theory of the general circulation because in the quasi-geostrophic model a number of assumptions are made about scales of motion which restrict in advance the possible types of solutions. Thus, most modern general circulation studies involve highly complex numerical models based on the primitive equations. The ultimate objective of these modeling efforts is to simulate the general circulation so faithfully that the climatological effects of any change in the external parameters can be accurately estimated. Present models are a long way from achieving this objective, but progress has been very rapid in recent years. The numerical simulation approach will be discussed in Section 11.6.

11.2   The Energy Cycle: A Quasi-Geostrophic Model

Perhaps the most important requirement for a theory of the general circulation is that it must provide a satisfactory explanation for the atmospheric energy cycle. It must, therefore, explain by what processes the atmosphere converts solar energy to kinetic energy in order to balance the kinetic energy losses by frictional dissipation. For a complete discussion of the energy cycle, it would be necessary to derive the energy equations starting with the primitive equations in spherical coordinates. However, in order to keep the discussion consistent with the general theme of this book we will here examine the energy cycle within the context of the quasi-geostrophic model. It turns out, in fact, that the basic energy cycle of the atmosphere can be described adequately by a properly formulated quasi-geostrophic model, at least in a qualitative sense. However, the quasi-geostrophic model in the form which we have previously used is not suitable for studying the atmospheric energy cycle because in that model we completely neglected the effects of friction and diabatic heating. For short-term forecasts, these omissions are perfectly justifiable since scaling arguments indicate that mildlatitude synoptic scale motions are nearly adiabatic and frictionless. However, the flow must ultimately depend on the energy sources and sinks, so that in a general circulation model, friction and diabatic heating must be included in some manner.

In Chapter 6 we showed that for synoptic scale motions the primary effect of surface friction is to produce a secondary circulation in which the vertical velocity at the top of the boundary layer is proportional to the vorticity at that level. Further, we showed that the divergent motions associated with this secondary circulation always act to reduce the relative vorticity of the flow, that is, to "spin-down" the motion. In addition to the energy lost to the surface through this indirect spin-down effect there is also an energy loss due to internal frictional dissipation. However, for simplicity we will include only the Ekman layer friction in our model. This is most easily done by letting the quasi-geostrophic model apply only above the Ekman layer and setting as the lower boundary condition the vertical velocity at the top of the Ekman layer given in (6.24). In terms of ω this level terrain condition becomes

$$\omega(p_0) = -\left[\rho g\left(\frac{K}{2f_0}\right)^{1/2} \nabla^2\psi\right]_{p_0} \tag{11.1}$$

where p_0 is the pressure at the top of the Ekman layer.

For purposes of the present discussion, it is not necessary to specify the form of diabatic heating explicitly. It is merely necessary to note from (1.20) that when diabatic heating is included the first law of thermodynamics becomes

$$c_p \frac{d\ln\theta}{dt} = \frac{dQ}{dt} \tag{11.2}$$

where dQ/dt is the heating rate per unit mass divided by the absolute temperature. In terms of the geostrophic streamfunction, $\psi = \Phi/f_0$, (11.2) can be written as

$$\frac{\partial}{\partial t}\left(\frac{\partial\psi}{\partial p}\right) = -\mathbf{V}_\psi \cdot \nabla\left(\frac{\partial\psi}{\partial p}\right) - \frac{\sigma}{f_0}\omega - \frac{\alpha}{f_0 c_p}\frac{dQ}{dt} \tag{11.3}$$

where

$$\mathbf{V}_\psi = \mathbf{k} \times \nabla \psi$$

and where we have used all the usual approximations of quasi-geostrophic theory, except the neglect of diabatic heating [compare with Eq. (8.13)].

The equations of our model are thus the quasi-geostrophic vorticity equation (8.12) and the thermodynamic energy equation (11.3). Using the condition

$$\nabla \cdot \mathbf{V}_\psi = 0$$

to write the advection terms in the so-called *flux form* by letting

$$\mathbf{V}_\psi \cdot \nabla \chi = \nabla \cdot (\mathbf{V}_\psi \chi)$$

where χ here stands for any field variable, we can write these equations as

$$\frac{\partial}{\partial t} \nabla^2 \psi = -\nabla \cdot [\mathbf{V}_\psi (\nabla^2 \psi + f)] + f_0 \frac{\partial \omega}{\partial p} \qquad (11.4)$$

$$\frac{\partial}{\partial t} \left(\frac{\partial \psi}{\partial p} \right) = -\nabla \cdot \left[\mathbf{V}_\psi \frac{\partial \psi}{\partial p} \right] - \frac{\sigma}{f_0} \omega - R \qquad (11.5)$$

where for simplicity we have set

$$R \equiv \frac{\alpha}{f_0 c_p} \frac{dQ}{dt}$$

We now assume that the atmosphere is confined to a zonal channel on the β plane with origin at 45° latitude and rigid walls at $y = \pm L$. This assumption may seem artificial, but it allows us to use cartesian coordinates and to avoid possible difficulties with the quasi-geostrophic model at the equator. The boundary conditions on the system (11.4) and (11.5) then become

$$\frac{\partial \psi}{\partial x} = 0 \qquad \text{at} \qquad y = \pm L \qquad (11.6)$$

together with $\omega = 0$ at $p = 0$ and ω given by (11.1) at $p = p_0$.

In order to study the zonally averaged circulation with this model, we divide the nondivergent motion into a zonally averaged portion denoted by an overbar plus the deviation from a zonal average denoted by a prime:

$$\psi = \bar{\psi} + \psi', \qquad \mathbf{V}_\psi = \overline{\mathbf{V}}_\psi + \mathbf{V}_\psi'$$

In the following discussion we will often refer to the zonally averaged part of the flow as the *mean flow* (it must be kept in mind that a zonal mean and not a time mean is implied). In addition the deviation from the mean will be referred to as the *eddy* motion (we will generally reserve the term *perturbation* for small deviations from the mean).

The primary goal of this section is to derive energy equations for both the mean and eddy components of the motion and to discuss the interactions between the eddy and mean flow energy. However, it is instructive to consider first the energy cycle of the complete flow without separating the field into mean and eddy parts. In a manner analogous to the discussion in Section 10.3.2, we can obtain the kinetic energy equation by multiplying (11.4) through by $-\psi$ and integrating over the volume of the channel to get

$$\int_0^{p_0} \int_A -\psi \frac{\partial}{\partial t} \nabla^2\psi \, dA \, dp = \int_0^{p_0} \int_A \psi \nabla \cdot [\mathbf{V}_\psi(\nabla^2\psi + f)] \, dA \, dp$$

$$- \int_0^{p_0} \int_A f_0\psi \frac{\partial \omega}{\partial p} \, dA \, dp \qquad (11.7)$$

The term on the left in (11.7) may be rewritten as

$$\int_A -\psi \frac{\partial}{\partial t} \nabla^2\psi \, dA = - \int_A \nabla \cdot \left(\psi \frac{\partial}{\partial t} \nabla\psi\right) dA + \int_A \nabla\psi \cdot \frac{\partial}{\partial t} \nabla\psi \, dA$$

Applying the divergence theorem together with the boundary condition (11.6) to the first term on the right above yields

$$\int_A \nabla \cdot \left(\psi \frac{\partial}{\partial t} \nabla\psi\right) dA = \int_l \left(\psi \frac{\partial}{\partial t} \nabla\psi\right) \cdot \mathbf{n} \, dl = 0$$

where l here designates the boundary lines at $y = \pm L$. Therefore,

$$\int_A -\psi \frac{\partial}{\partial t} \nabla^2\psi \, dA = \int_A \frac{\partial}{\partial t} \left(\frac{\nabla\psi \cdot \nabla\psi}{2}\right) dA$$

$$= \int_A \frac{\partial}{\partial t} \left(\frac{\mathbf{V}_\psi \cdot \mathbf{V}_\psi}{2}\right) dA \qquad (11.8)$$

The first term on the right in (11.7) may also be shown to equal zero with the aid of the divergence theorem. Thus, we write

$$\int_A \psi \nabla \cdot [\mathbf{V}_\psi(\nabla^2\psi + f)] \, dA = \int_A \nabla \cdot [\psi \mathbf{V}_\psi(\nabla^2\psi + f)] \, dA$$

$$- \int_A (\nabla^2\psi + f)\mathbf{V}_\psi \cdot \nabla\psi \, dA \qquad (11.9)$$

Applying the divergence theorem we find that the first term on the right in (11.9) is zero. The second term on the right is also zero because

$$\mathbf{V}_\psi \cdot \nabla\psi = (\mathbf{k} \times \nabla\psi) \cdot \nabla\psi \equiv 0$$

Therefore, there is no contribution to the change of average kinetic energy in a closed region due to horizontal advection of kinetic energy.

The final term in (11.7) can be expressed in a more revealing form if we integrate by parts in the vertical:

$$-f_0 \int_0^{p_0} \psi \frac{\partial \omega}{\partial p} \, dp = f_0 \int_0^{p_0} \omega \frac{\partial \psi}{\partial p} \, dp - f_0 [\psi \omega]_{p_0} \tag{11.10}$$

where we have applied the boundary condition $\omega = 0$ at $p = 0$. Substituting from the boundary condition (11.1) into the last term of (11.10) and integrating over the area of the channel we obtain

$$-f_0 \int_A [\psi \omega]_{p_0} \, dA = \int_A \rho g \left(\frac{K f_0}{2}\right)^{1/2} [\psi \nabla^2 \psi]_{p_0} \, dA$$

$$= -\int_A \left[\rho g \left(\frac{K f_0}{2}\right)^{1/2} \frac{(\nabla \psi)^2}{2} \right]_{p_0} dA \equiv -\varepsilon \tag{11.11}$$

Thus, ε is the rate of frictional dissipation due to the surface stress. It is clear from the form of the integrand in (11.11) that ε will always be positive. Substituting from (11.8)–(11.11) back into (11.7) we obtain finally the quasi-geostrophic kinetic energy equation

$$\frac{d}{dt} \int_0^{p_0} \int_A \frac{(\nabla \psi)^2}{2} \, dA \, dp = \int_0^{p_0} \int_A f_0 \omega \frac{\partial \psi}{\partial p} \, dA \, dp - \varepsilon \tag{11.12}$$

To obtain an equation for the rate of change of available potential energy, we multiply (11.5) through by

$$\frac{f_0^2}{\sigma} \frac{\partial \psi}{\partial p}$$

and integrate over the entire channel. In this case, the first term on the right may be expressed as

$$-\frac{f_0^2}{\sigma} \left(\frac{\partial \psi}{\partial p}\right) \mathbf{V} \cdot \left[\mathbf{V}_\psi \frac{\partial \psi}{\partial p} \right] = -\frac{f_0^2}{2\sigma} \mathbf{V} \cdot \left[\mathbf{V}_\psi \left(\frac{\partial \psi}{\partial p}\right)^2 \right] \tag{11.13}$$

where we have used the relation $\mathbf{V} \cdot \mathbf{V}_\psi = 0$ to eliminate several terms. Clearly when we integrate over the area of the channel and apply the divergence theorem the expression in (11.13) will give zero contribution. Thus, the available potential energy equation becomes

$$\frac{d}{dt} \int_0^{p_0} \int_A \frac{f_0^2}{2\sigma} \left(\frac{\partial \psi}{\partial p}\right)^2 dA \, dp = -\int_0^{p_0} \int_A f_0 \omega \frac{\partial \psi}{\partial p} \, dA \, dp$$

$$-\int_0^{p_0} \int_A \frac{f_0^2}{\sigma} R \frac{\partial \psi}{\partial p} \, dA \, dp \tag{11.14}$$

It is convenient at this point to introduce some new notation by defining the total kinetic energy K and the available potential energy P as

$$K \equiv \int_0^{p_0} \int_A \frac{(\nabla\psi)^2}{2}\, dA\, dp$$

$$P \equiv \int_0^{p_0} \int_A \frac{f_0{}^2}{2\sigma} \left(\frac{\partial\psi}{\partial p}\right)^2 dA\, dp$$

further, we let

$$\{P \cdot K\} \equiv \int_0^{p_0} \int_A f_0\omega \frac{\partial\psi}{\partial p}\, dA\, dp$$

$$\{R \cdot P\} \equiv -\int_0^{p_0} \int_A \frac{f_0{}^2}{\sigma} R \frac{\partial\psi}{\partial p}\, dA\, dp$$

Then (11.12) and (11.14) can be written symbolically as

$$\frac{dK}{dt} = \{P \cdot K\} - \varepsilon \tag{11.15}$$

$$\frac{dP}{dt} = -\{P \cdot K\} + \{R \cdot P\} \tag{11.16}$$

From the form of (11.15) and (11.16) it is clear that $\{P \cdot K\}$ represents the conversion of available potential energy to kinetic energy due to the correlation of temperature and vertical motion, while $\{R \cdot P\}$ represents the generation of available potential energy due to the correlation of temperature and diabatic heating. Adding (11.15) and (11.16), we obtain the total energy equation for the quasi-geostrophc system

$$\frac{d}{dt}(P + K) = \{R \cdot P\} - \varepsilon \tag{11.17}$$

In the long-term mean, $P + K$ must be constant so that in this model the generation of available potential energy by differential heating must just balance the kinetic energy loss due to surface friction.

In the above discussion we have illustrated the general technique of deriving energy equations for the quasi-geostrophic model. We now discuss the equations which arise when the energy is partitioned between its zonal mean and eddy portions.

Dividing the variables in (11.4) into their mean and eddy parts and taking the zonal mean we obtain the mean vorticity equation

$$\frac{\partial}{\partial t}\left(\frac{\partial^2\overline{\psi}}{\partial y^2}\right) = \frac{\partial^2 M}{\partial y^2} + f_0 \frac{\partial\overline{\omega}}{\partial p} \tag{11.18}$$

Here, letting angle brackets as well as overbars denote zonally averaged quantities

$$\frac{\partial^2 M}{\partial y^2} \equiv -\langle \mathbf{V} \cdot [\overline{\mathbf{V}}_\psi + \mathbf{V}_\psi{}')(\nabla^2 \overline{\psi} + \nabla^2 \psi' + f)]\rangle$$

$$= -\frac{\partial}{\partial y} \left\langle \left[\frac{\partial \psi'}{\partial x} \nabla^2 \psi' \right] \right\rangle$$

where we have used the facts that $\partial \langle [\ ] \rangle / \partial x = 0$ for any function and $\overline{\mathbf{V}}_\psi{}' = 0$. The quantity M may be given a simple physical interpretation as follows:

$$\frac{\partial M}{\partial y} = -\left\langle \frac{\partial \psi'}{\partial x} \nabla^2 \psi' \right\rangle = -\left\langle \frac{\partial \psi'}{\partial x} \frac{\partial^2 \psi'}{\partial x^2} + \frac{\partial \psi'}{\partial x} \frac{\partial^2 \psi'}{\partial y^2} \right\rangle$$

$$= -\frac{\partial}{\partial y} \left\langle \left(\frac{\partial \psi'}{\partial x} \frac{\partial \psi'}{\partial y} \right) \right\rangle = \frac{\partial}{\partial y} (u'v') \qquad (11.19)$$

where we have again used $\partial \langle [\ ] \rangle / \partial x = 0$. Thus, $M = \langle u'v' \rangle$ is just the correlation between the eddy zonal and meridional velocity components averaged around a latitude circle.

Subtracting (11.18) from (11.4) we obtain the eddy vorticity equation

$$\frac{\partial}{\partial t} \nabla^2 \psi' = -\mathbf{V} \cdot [(\overline{\mathbf{V}}_\psi + \mathbf{V}_\psi{}')(\nabla^2 \overline{\psi} + \nabla^2 \psi' + f)] + f_0 \frac{\partial \omega'}{\partial p} - \frac{\partial^2 M}{\partial y^2} \quad (11.20)$$

We obtain the mean kinetic energy equation by multiplying (11.18) by $-\overline{\psi}$ and integrating over the entire volume. After some manipulation we obain

$$\frac{d\overline{K}}{dt} = \{K' \cdot \overline{K}\} + \{\overline{P} \cdot \overline{K}\} - \bar{\varepsilon} \qquad (11.21)$$

where

$$\overline{K} \equiv \int_0^{p_o} \int_A \frac{(\nabla \overline{\psi})^2}{2} \, dA \, dp$$

is the mean kinetic energy,

$$\{K' \cdot \overline{K}\} \equiv \int_0^{p_o} \int_A \overline{\psi} \frac{\partial^2 M}{\partial y^2} \, dA \, dp$$

is the conversion of eddy kinetic energy to mean kinetic energy,

$$\{\overline{P} \cdot \overline{K}\} \equiv \int_0^{p_o} \int_A f_0 \overline{\omega} \frac{\partial \overline{\psi}}{\partial p} \, dA \, dp$$

is the conversion of mean available potential energy to mean kinetic energy and

$$\bar{\varepsilon} \equiv \int_A \left[\rho g \left(\frac{Kf_0}{2} \right)^{1/2} \frac{(\nabla\bar{\psi})^2}{2} \right] dA$$

is the surface frictional dissipation for the mean wind. Similarly, multiplying through (11.20) by ψ' and integrating over the entire volume, we obtain

$$\frac{dK'}{dt} = -\{K' \cdot \bar{K}\} + \{P' \cdot K'\} - \varepsilon' \tag{11.22}$$

where

$$K' \equiv \int_0^{p_0} \int_A \frac{(\nabla\psi')^2}{2} \, dA \, dp$$

is the eddy kinetic energy,

$$\{P' \cdot K'\} = \int_0^{p_0} \int_A f_0 \, \omega' \frac{\partial \psi'}{\partial p} \, dA \, dp$$

is the conversion of eddy available potential energy to eddy kinetic energy, and,

$$\varepsilon' \equiv \int_A \left[\rho g \left(\frac{Kf_0}{2} \right)^{1/2} \frac{(\nabla\psi')^2}{2} \right] dA$$

is the eddy surface frictional dissipation.

We now turn to the thermodynamic energy equation. Again dividing the variables into their mean and eddy parts, we obtain from (11.5) the mean equation

$$\frac{\partial}{\partial t} \left(\frac{\partial \bar{\psi}}{\partial p} \right) = + \frac{\partial B}{\partial y} - \frac{\sigma}{f_0} \bar{\omega} - \bar{R} \tag{11.23}$$

where

$$B \equiv - \left\langle \frac{\partial \psi'}{\partial x} \frac{\partial \psi'}{\partial p} \right\rangle$$

is proportional to the correlation between temperature and meridional velocity averaged around a latitude circle.

Subtracting (11.23) from (11.5) we obtain the eddy thermodynamic energy equation

$$\frac{\partial}{\partial t} \left(\frac{\partial \psi'}{\partial p} \right) = -\nabla \cdot \left[(\bar{\mathbf{V}}_\psi + \mathbf{V}_{\psi'}) \cdot \nabla \left(\frac{\partial \bar{\psi}}{\partial p} + \frac{\partial \psi'}{\partial p} \right) \right] - \frac{\sigma}{f_0} \omega' - R' - \frac{\partial B}{\partial y} \tag{11.24}$$

Multiplying (11.23) and (11.24) through by

$$\frac{f_0^2}{\sigma} \frac{\partial \bar{\psi}}{\partial p} \quad \text{and} \quad \frac{f_0^2}{\sigma} \frac{\partial \psi'}{\partial p}$$

respectively, we obtain the mean available potential energy equation

$$\frac{d\bar{P}}{dt} = -\{\bar{P} \cdot P'\} - \{\bar{P} \cdot \bar{K}\} + \{\bar{R} \cdot \bar{P}\} \qquad (11.25)$$

and the eddy available potential energy equation

$$\frac{dP'}{dt} = +\{\bar{P} \cdot P'\} - \{P' \cdot K'\} - \{P' \cdot R'\} \qquad (11.26)$$

where

$$\bar{P} \equiv \int_0^{p_0} \int_A \frac{f_0^2}{2\sigma} \left(\frac{\partial \bar{\psi}}{\partial p}\right)^2 dA\, dp$$

is the mean available potential energy.

$$P' = \int_0^{p_0} \int_A \frac{f_0^2}{2\sigma} \left(\frac{\partial \psi'}{\partial p}\right)^2 dA\, dp$$

is the eddy available potential energy,

$$\{\bar{P} \cdot P'\} \equiv -\int_0^{p_0} \int_A \frac{f_0^2}{\sigma} \frac{\partial \bar{\psi}}{\partial p} \frac{\partial B}{\partial y} dA\, dp$$

is the conversion of mean available potential energy to eddy available potential energy,

$$\{\bar{R} \cdot \bar{P}\} \equiv -\int_0^{p_0} \int_A \frac{f_0^2}{\sigma} \bar{R} \frac{\partial \bar{\psi}}{\partial p} dA\, dp$$

is the generation of mean available potential energy by the mean heating, and

$$\{P' \cdot R'\} \equiv \int_0^{p_0} \int_A \frac{f_0^2}{\sigma} R' \frac{\partial \psi'}{\partial p} dA\, dp$$

is the radiative dissipation in the eddies.

Combining (11.21), (11.22), (11.25), and (11.26), we obtain the total energy equation

$$\frac{d}{dt}(\bar{K} + K' + \bar{P} + P') = \{\bar{R} \cdot \bar{P}\} - \{P' \cdot R'\} - \bar{\varepsilon} - \varepsilon' \qquad (11.27)$$

which should be compared with (11.17). Thus, in the long-term time mean

$$\{\bar{R} \cdot \bar{P}\} = \{P' \cdot R'\} + \bar{\varepsilon} + \varepsilon' \qquad (11.28)$$

Now $\{P' \cdot R'\}$ will be positive provided that the eddy diabatic heating and eddy temperatures are positively correlated. For a dry atmosphere in which

R' is due entirely to radiation and diffusion, $\{P' \cdot R'\}$ will generally be positive because the infrared radiation of the atmosphere to space increases with increasing temperature. Thus, since $\bar{\varepsilon}$ and ε' are invariably positive we see that the three terms on the right in (11.28) represent energy losses which must in the long-term mean be balanced by the generation of zonal mean potential energy by the zonal mean heating.

The equations (11.21), (11.22), (11.25), and (11.26) together provide a complete description of the atmospheric energy cycle, within the limits of the quasi-geostrophic theory. The content of these equations is summarized by means of the energy cycle diagram shown in Fig. 11.1. In this diagram the squares represent reservoirs of energy and the arrows indicate sources, sinks, and conversions of energy. The observed direction of the conversion terms in the troposphere for the Northern Hemisphere annual mean is indicated by the arrows It should be emphasized that the direction of the various conversions cannot be theoretically deduced by reference to the energy equations alone. However, the observed energy cycle as summarized in Fig. 11.1 suggests the following qualitative picture:

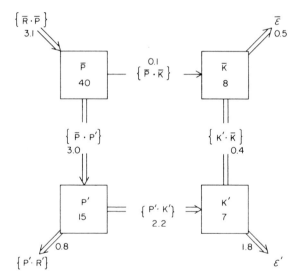

Fig. 11.1   The atmospheric energy cycle. Numbers in the squares give observed annual mean Northern Hemisphere energy values in units of 10^5 joules m^{-2}, numbers next to the arrows give rates of conversion, generation and dissipation in W m^{-2}. (Data is from Oort, 1964.)

1. The zonal mean radiative heating generates mean zonal available potential energy through a net heating of the tropics and cooling of the polar regions.

2. Baroclinic eddies transport warm air northward, cold air southward, and transform the mean available potential energy to eddy available potential energy.

3. At the same time eddy available potential energy is transformed into eddy kinetic energy by the vertical motions in the eddies.

4. The zonal kinetic energy is maintained primarily by the conversion from eddy kinetic energy due to the correlation $\langle u'v' \rangle$. This will be discussed further in the next section.

5. The energy is dissipated by surface and internal friction in the eddies and mean flow plus radiative damping in the eddies. The eddies tend to have higher vorticity than the mean flow at the top of the Ekman layer, hence much more of the surface dissipation is in the eddies than in the mean flow despite the fact that the mean flow kinetic energy is larger than the eddy kinetic energy.

In summary, the observed atmospheric energy cycle is consistent with the notion that eddies which result from the baroclinic instability of the mean flow are to a large extent responsible for the energy exchange in the atmosphere. It is through the eddy motions that the kinetic energy lost through friction is replaced, and it is the eddies which are primarily responsible for the poleward heat transport to balance the radiation deficit in the polar regions. In addition to the transient baroclinic eddies, forced stationary orographic waves and free Rossby waves may also contribute substantially to the poleward heat flux. The direct conversion of mean available potential energy to mean kinetic energy by symmetric overturning is, on the other hand, small and negative in middle latitudes, but positive in the tropics where it plays an important role in the maintenance of the mean Hadley circulation.

11.3 The Momentum Budget

In the previous section we saw that the observed zonal mean kinetic energy was maintained primarily by the eddies through the conversion $\{K' \cdot \overline{K}\}$. This process can alternatively be interpreted in terms of the atmospheric momentum budget. The angular momentum of the earth and atmosphere combined must remain constant in time except for the small effects of tidal friction. Since the average rotation rate of the earth is itself observed to be very close to consant, the atmosphere must also on the average conserve its angular momentum. Since the atmosphere gains angular momentum from the earth in the tropics where the surface winds are easterly and gives up angular momentum to the earth in middle latitudes where the surface winds are westerly, there must be a net poleward transport of angular momentum within the atmosphere, otherwise the torque due to surface friction would decelerate both the easterlies and westerlies. Furthermore, the angular momentum given by the earth to the atmosphere in the belt of easterlies must just balance the angular momentum given up to the earth in the westerlies if the angular momentum of the atmosphere is to remain constant.

In the equatorial regions the poleward momentum transport is divided between the advection of absolute angular momentum by the poleward flow in the axially symmetric Hadley circulation and transport by eddies. However, in the middle latitudes the zonal mean meridional velocity $\bar{v}$ is much too small to account for a significant fraction of the required transport. Thus, it must be primarily the eddy motions which transport momentum northward in the middle latitudes and the angular momentum budget of the atmosphere must be qualitatively as shown in Fig. 11.2.

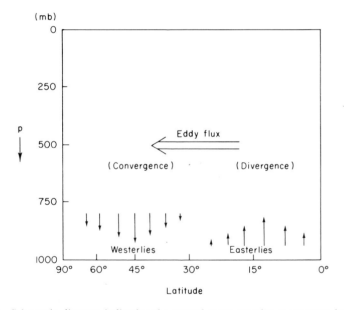

Fig. 11.2 Schematic diagram indicating the annual mean angular momentum budget in the atmosphere.

Observations indicate a maximum poleward flux of angular momentum at about 30°N and a maximum horizontal flux convergence at about 45°N. This maximum in the flux convergence is a reflection of the strong $\{K' \cdot \bar{K}\}$ energy conversion in the upper level westerlies and is the mechanism whereby the atmosphere can maintain a positive zonal wind in the middle latitudes despite the momentum lost to the surface.

Since absolute angular momentum is conserved during a latitudinal displacement (neglecting frictional effects), it is convenient to analyze the momentum budget in terms of the absolute angular momentum. The absolute angular momentum per unit mass of atmosphere is

$$\mu = (\Omega a \cos \phi + u)a \cos \phi$$

The absolute angular momentum of an individual air parcel can be changed only by the torques, due to the zonal pressure gradient and the viscous stresses. Thus, from Newton's second law in its angular-momentum form

$$\frac{d\mu}{dt} = -\frac{a \cos \phi}{\rho} \frac{\partial p}{\partial x} - \frac{a \cos \phi}{\rho} \frac{\partial \tau}{\partial z} \qquad (11.29)$$

where it is assumed that horizontal viscous stresses are negligible compared to the vertical stress.

Multiplying (11.29) through by ρ and adding the result to the continuity equation multiplied by μ,

$$\mu \frac{d\rho}{dt} + \mu\rho \, \mathbf{V} \cdot \mathbf{V} = 0$$

we obtain

$$\frac{\partial}{\partial t}(\rho\mu) = -\mathbf{V} \cdot (\rho\mu \, \mathbf{V}) - a \cos \phi \frac{\partial p}{\partial x} - a \cos \phi \frac{\partial \tau}{\partial z} \qquad (11.30)$$

which is an expression for the rate of change of absolute angular momentum per unit volume. Integrating (11.30) in height and zonally around a latitude circle we obtain the rate of change of absolute angular momentum for a zonal ring of air of unit meridional width at the latitude ϕ:

$$\frac{d}{dt} \int_0^\infty \int_0^{2\pi} \rho\mu a \cos \phi \, d\lambda \, dz = -\int_0^\infty \int_0^{2\pi} \frac{\partial}{\partial y}(\rho\mu v)a \cos \phi \, d\lambda \, dz$$

$$- \int_0^\infty \int_0^{2\pi} (a \cos \phi)^2 \frac{\partial p}{\partial x} \, d\lambda \, dz$$

$$- (a \cos \phi)^2 \int_0^{2\pi} \tau_{z=0} \, d\lambda \qquad (11.31)$$

where $\tau_{z=0}$ designates the viscous stress at the ground.

In the long-term mean the time derivative in (11.31) is zero so that the three terms on the right must balance. The first term on the right represents the convergence of absolute angular momentum into the ring due to the meridional transport. This term may be best interpreted physically by dividing the flow into zonal mean and eddy components as in Section 11.2. Thus, we let

$$\mu = \bar{\mu} + \mu' = (\Omega a \cos \phi + \bar{u} + u')a \cos \phi$$
$$v = \bar{v} + v'$$

in which case

$$\langle \rho \mu v \rangle = \rho[\Omega a \cos \phi \bar{v} + \bar{u}\bar{v} + \langle u'v' \rangle]a \cos \phi \qquad (11.32)$$

The three parts of the meridional momentum flux given in (11.32) are called the Ω-momentum flux, the drift, and the eddy momentum flux, respectively. Since $\bar{u} \ll \Omega a \cos \phi$ the second term (the drift) is usually neglected. Further, if we integrate in height and observe that in the long-term mean mass continuity requires

$$\int_0^\infty \rho \bar{v} \, dz = 0$$

we get approximately[1]

$$\int_0^\infty \int_0^{2\pi} a \cos \phi \frac{\partial}{\partial y} (\rho \mu v) \, d\lambda \, dz \simeq \int_0^\infty \rho a^2 \cos \phi \frac{\partial}{\partial y} (M \cos \phi) \, dz$$

where we have neglected the variation of ρ in integrating zonally, and $M = \langle u'v' \rangle$ is the correlation between the eddy zonal and meridional velocity components averaged around a latitude circle [see (11.19)].

According to the scheme shown in Fig. 11.2, M should be positive and decreasing with latitude in the belt of westerlies. If M is to be positive, the eddies must be asymmetric in the horizonal plane with the trough and ridge axes tilting as indicated in Fig. 11.3. When the troughs and ridges on the average have a southwest to northeast tilt the zonal flow will be larger than average ($u' > 0$) when the meridional flow is poleward ($v' > 0$) and zonal flow will be less than the average ($u' > 0$) for equatorward flow. Thus, $\langle u'v' \rangle > 0$ and the eddies will systematically transport positive zonal momentum poleward. The quantity M represents the rate of eddy momentum transport per unit area, or momentum flux, so that $\partial(M \cos \phi)/\partial y$ indeed represents the convergence of the eddy momentum flux.

Referring back to (11.31), we see that the eddy momentum flux convergence must balance the sum of the pressure torque and frictional torque. The

---

[1] The approximation here arises not only from the neglect of the drift term but also from the fact that to be exact the Ω-momentum term should be $\Omega(a + z) \cos \phi \, \bar{v}$ in which case the vertical integral of the Ω-momentum flux would not generally be zero since

$$\int_0^\infty z \rho \bar{v} \, dz$$

is usually not zero.

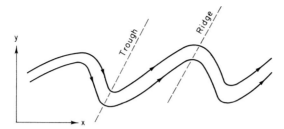

Fig. 11.3 Schematic streamlines for a positive eddy momentum flux.

second term on the right in (11.31), the pressure torque, would be zero for a zonal ring of air on a flat earth since in that case

$$\int_0^{2\pi} a \cos \phi \, \frac{\partial p}{\partial x} \, d\lambda = \oint dp = 0$$

However, in the presence of mountains there can be a net contribution from the pressure torque if there are pressure differences between the western and eastern sides of the mountains. Observations indicate that the pressure torque is an important angular momentum sink in the middle latitudes of the Northern Hemisphere due to the lower pressure on the eastern sides of mountains (the commonly observed "lee side trough").

The final term in (11.31) represents the angular-momentum exchange with the earth's surface due to the surface frictional torque. Estimates of the torques due to surface friction and the mountain effect have been attempted by several investigators despite the lack of adequate global data coverage and the difficulty in estimating the surface stress. One such estimate of the latitudinal distributon of the eastward torque exerted by the earth on the atmosphere is shown in Fig. 11.4a. The mountain torque has been included implicitly in this estimate by raising the measured midlatitude surface torque by 40 percent so that the net latitudinally integrated torque balanced to zero. The required northward flux of angular momentum to balance the estimated surface torque is also indicated in Fig. 11.4a. This flux can also be directly estimated from wind data using (11.32). In Fig. 11.4b an estimate based on the observed winds is compared with the northward angular momentum flux required to balance the torques shown in Fig. 11.4a. Considering the many uncertainties in these measurements, the agreement is remarkable. It should also be mentioned here that except for the belt within 10° of the equator, almost all of the northward flux is due to the eddy flux term $\langle u'v' \rangle$. Thus, the momentum budget and the energy cycle both depend critically on the transports by the eddies which result in part from the baroclinic instability of the basic zonal current.

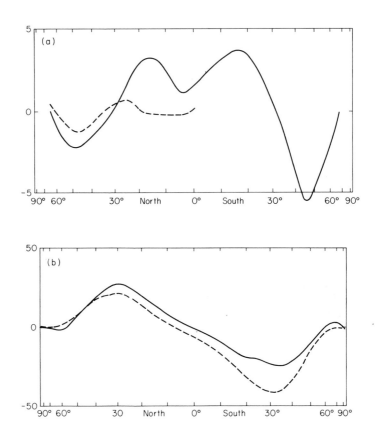

Fig. 11.4 (a) Average eastward torque exerted on the atmosphere by surface friction (solid curve) and by mountains in the northern hemisphere (dashed curve) in units of 10^8 gm sec^{-2}. (b) The observed transport of angular momentum in units of 10^{25} gm cm^2 sec^{-2} (solid curve) and the required transport (dashed curve) as given by the observed surface torques. (After Lorenz, 1967).

11.4   The Dynamics of Zonally Symmetric Circulations

In order to study the distribution of the mean zonal wind quantitatively, it is necessary to derive an equation which relates the zonal mean momentum to the sources, sinks, and transports of momentum. In this section we will show that zonally symmetric motions may be understood as a special case of the quasi-geostrophic theory, and that the vertical flow associated with an axially symmetric vortex is dynamically analogous to the vertical circulation in a baroclinic wave.

In the quasi-geostrophic model which we used to discuss the energy cycle in Section 11.2 internal friction and radiative dissipation were neglected in the equations for the mean zonal motion. Actually, since the mean zonal circulation changes rather slowly in time, internal dissipation should not be neglected. Internal damping is essential as a mechanism to balance the eddy momentum and heat sources in the long-term mean. A qualitative understanding of the role of internal dissipation may be obtained by representing this process very simply as a linear drag (Rayleigh friction) in the momentum equation and a linear thermal damping (Newtonian cooling) in the thermodynamic energy equation. For symmetry, we assume that the rate coefficient d is the same for both of these processes. The zonal mean vorticity equation (11.18) and zonal mean thermodynamic energy equation (11.23) then become

$$\left(\frac{\partial}{\partial t} + d\right)\left(\frac{\partial^2 \bar{\psi}}{\partial y^2}\right) = \frac{\partial^2 M}{\partial y^2} + f_0 \frac{\partial \bar{\omega}}{\partial p} \tag{11.33}$$

$$\left(\frac{\partial}{\partial t} + d\right)\left(\frac{\partial \bar{\psi}}{\partial p}\right) = \frac{\partial B}{\partial y} - \frac{\sigma}{f_0}\bar{\omega} - \bar{R} \tag{11.34}$$

This is a closed set in the variables $\bar{\psi}$ and $\bar{\omega}$ provided that the momentum flux M, the heat flux B, and the diabatic heating rate $\bar{R}$ are known.

If we eliminate $\bar{\omega}$ between (11.33) and (11.34) we obtain a *potential vorticity equation* for the mean flow:

$$\left(\frac{\partial}{\partial t} + d\right)\bar{q} = \frac{\partial^2 M}{\partial y^2} + \frac{f_0^2}{\sigma}\frac{\partial^2 B}{\partial y \partial p} - \frac{f_0^2}{\sigma}\frac{\partial \bar{R}}{\partial p} \tag{11.35}$$

where

$$\bar{q} = \frac{\partial^2 \bar{\psi}}{\partial y^2} + \frac{f_0^2}{\sigma}\frac{\partial^2 \bar{\psi}}{\partial p^2}$$

Now from the definition of B we can write, again using angle brackets instead of overbars to indicate averaged quantities

$$\frac{\partial B}{\partial p} = -\frac{\partial}{\partial p}\left\langle\left(\frac{\partial \psi'}{\partial x}\frac{\partial \psi'}{\partial p}\right)\right\rangle = -\left\langle\frac{\partial \psi'}{\partial x}\frac{\partial^2 \psi'}{\partial p^2}\right\rangle$$

where we have used the fact that

$$\left\langle\left(\frac{\partial \psi'}{\partial p}\right)\frac{\partial}{\partial x}\left(\frac{\partial \psi'}{\partial p}\right)\right\rangle = \frac{1}{2}\left\langle\frac{\partial}{\partial x}\left(\frac{\partial \psi'}{\partial p}\right)^2\right\rangle = 0$$

Recalling that

$$-\frac{\partial M}{\partial y} = \left\langle\left(\frac{\partial \psi'}{\partial x}\nabla^2 \psi'\right)\right\rangle$$

we can now combine the terms in (11.35) involving B and M according to

$$\frac{\partial^2 M}{\partial y^2} + \frac{f_0^2}{\sigma} \frac{\partial}{\partial y} \left(\frac{\partial B}{\partial p}\right) = -\frac{\partial}{\partial y} \langle (q'v') \rangle$$

where

$$v' = \frac{\partial \psi'}{\partial x} \quad \text{and} \quad q' \equiv \nabla^2 \psi' + \frac{f_0^2}{\sigma} \frac{\partial^2 \psi'}{\partial p^2}$$

is the eddy potential vorticity.

Therefore, in the absence of diabatic heating and internal damping (11.35) can be written simply as

$$\frac{\partial \bar{q}}{\partial t} = -\frac{\partial}{\partial y} \langle (v'q') \rangle$$

which states that the zonal mean potential vorticity is changed only if there is potential vorticity transport by the eddies.

For physical interpretation of the processes which maintain the zonal mean circulation, it is easier to work with the zonal momentum equation rather than the potential vorticity equation. Using the zonally averaged continuity equation

$$\frac{\partial \bar{v}}{\partial y} = -\frac{\partial \bar{\omega}}{\partial p}$$

where $\bar{v}$ is the zonally averaged meridional velocity, we may eliminate $\partial \bar{\omega}/\partial p$ from (11.33) and immediately integrate once with respect to y to obtain

$$\left(\frac{\partial}{\partial t} + d\right)\left(-\frac{\partial \bar{\psi}}{\partial y}\right) = -\frac{\partial M}{\partial y} + f\bar{v} \tag{11.36}$$

We next define a stream function X for the mean meridional circulation by letting

$$\bar{\omega} = \frac{\partial X}{\partial y} \quad \text{and} \quad \bar{v} = -\frac{\partial X}{\partial p}$$

The continuity equation is then identically satisfied. Differentiating (11.34) with respect to y, (11.36) with respect to p, and adding the resulting equations, we obtain after representing $\bar{\omega}$ and $\bar{v}$ in terms of X

$$\left(\frac{\partial^2}{\partial y^2} + \frac{f_0^2}{\sigma} \frac{\partial^2}{\partial p^2}\right)X = -\frac{f_0}{\sigma} \frac{\partial}{\partial p}\left(\frac{\partial M}{\partial y}\right) + \frac{\partial^2 B}{\partial y^2} - \frac{\partial \bar{R}}{\partial y} \tag{11.37}$$

Equation (11.37) is a diagnostic equation for the mean meridional circulation which is exactly analogous to the omega equation (7.21) for quasi-geostrophic disturbances. The terms in (11.37) thus have physical interpretations similar to those of the analogous terms in the omega equation. The term on the left, for example, is identical to the term on the left in the omega equation but with ω replaced by X. By the arguments presented in Section 7.3.2 we can thus write

$$\left(\frac{\partial^2}{\partial y^2} + \frac{f_0^2}{\sigma}\frac{\partial^2}{\partial p^2}\right)X \propto -X$$

Assuming that there is no circulation across the equator, X must be zero at the equator and at the pole; otherwise, the average value of $\bar{\omega}$ integrated over the hemisphere would not be zero since

$$\bar{\omega}\,dy = dX$$

Thus, as indicated in Fig. 11.5, in the Northern Hemisphere a negative X

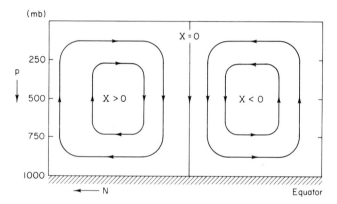

Fig. 11.5 Relation of the meridional streamfunction X to the vertical and meridional velocities.

implies a *direct* meridional circulation with rising motion to the south, poleward motion above, sinking motion to the north, and equatorward motion below the circulation center. Conversely, positive X implies an indirect meridional cell. Thus, by evaluating the forcing terms on the right in (11.37) we can immediately deduce the sense of the mean meridional circulation since

$$X \propto \frac{f_0}{\sigma}\frac{\partial}{\partial p}\left(\frac{\partial M}{\partial y}\right) - \frac{\partial^2 B}{\partial y^2} + \frac{\partial \bar{R}}{\partial y} \qquad (11.38)$$

The first term on the right in (11.38) is proportional to the vertical gradient of the horizontal eddy momentum flux convergence. This momentum term plays the same role in forcing a mean meridional circulation as the differential vorticity advection term of the omega equation plays in forcing vertical motions in a baroclinic disturbance. To interpret the momentum forcing physically, we suppose as shown in Fig. 11.6 that the momentum flux convergence is positive and increasing with height. Then

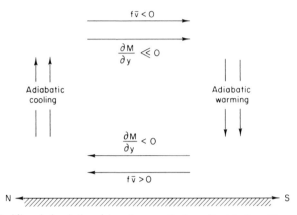

Fig. 11.6   Meridional circulation driven by a vertical gradient in the eddy momentum flux convergence.

$$\frac{\partial}{\partial p}\left(\frac{\partial M}{\partial y}\right) > 0$$

which implies that X is also greater than zero. But from (11.36) we see that unless this term is balanced everywhere by either the linear drag or the Coriolis torque, the zonal mean thermal wind must be increasing in time, that is,

$$\frac{\partial}{\partial t}\left(-\frac{\partial \bar{u}}{\partial p}\right) > 0$$

And in order to maintain geostrophic and hydrostatic balance the latitudinal temperature gradient must be simultaneously increasing. However, in the absence of eddy heat flux and diabatic heating the increased temperature gradient can only be provided by an indirect meridional circulation as shown in Fig. 11.5 with rising and adiabatic cooling north of the momentum source region and sinking and adiabatic warming south of the momentum source. Thus the circulation implied by (11.38) is consistent with the dynamical constraints of geostrophic and hydrostatic balance. In this case the meridional velocities required by continuity create Coriolis torques which tend to offset partially the effects of the momentum source by contributing easterly acceleration above and a westerly acceleration below. Thus, the meridional circulation here plays a role analogous to the divergent motions in baroclinic waves which, as we found in Section 3.7.2, tend to oppose the effects of differential vorticity advection.

Turning now to the second term on the right in (11.38) we first recall that

$$B = - \left\langle \frac{\partial \psi'}{\partial x} \frac{\partial \psi'}{\partial p} \right\rangle \propto \langle v'T' \rangle$$

is a measure of the northward eddy heat flux. Thus, $\partial^2 B/\partial y^2$ in (11.38) is analogous to the term involving the Laplacian of the temperature advection in the omega equation. At the latitude of maximum eddy heat flux B will be a maximum and $\partial^2 B/\partial y^2 < 0$, so that from (11.38) $X > 0$ and there will be an indirect meridional circulation centered about that latitude. The physical requirement for such an indirect meridional circulation can again be understood in terms of the need to maintain geostrophic and hydrostatic balance. North of the latitude where B is a maximum there is a convergence of eddy heat flux, while south of that latitude there is a divergence. Thus, the eddy heat transport will tend to reduce the pole to equator mean temperature gradient. If the mean zonal flow is to remain geostrophic, the thermal wind must then also decrease. In the absence of eddy momentum transport, this decrease in the thermal wind can only be produced by the Coriolis torques due to a mean meridional circulation as shown in Fig. 11.7. At the same time, it is not surprising to find that the vertical mean motions required by continuity oppose the temperature changes due to the eddy heat flux through adiabatic warming in the region of eddy heat flux divergence, and adiabatic cooling in the region of eddy heat flux convergence.

The final term in (11.38) is simply the zonal mean differential heating. In the usual case for the troposphere there is net diabatic heating in the tropics and cooling in the polar regions. Thus, $\partial \bar{R}/\partial y < 0$ so that $X < 0$ and the

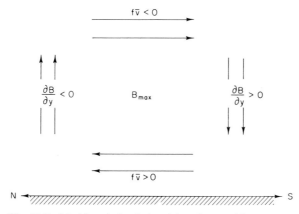

Fig. 11.7 Meridional circulation driven by an eddy heat flux.

diabatic heating drives a direct meridional circulation. Again this circulation may be understood in terms of the hydrostatic and geostrophic constraints. The differential heating increases the pole-to-equator temperature gradient which requires an increase in the zonal thermal wind. In the absence of eddy momentum sources the increased thermal wind can only be produced by the Coriolis torques of a direct circulation with poleward motion aloft and equatorward motion near the surface. Again, the adiabatic temperature changes due to the vertical motions associated with the direct cell act to oppose the diabatic heating.

The direct circulation which we have just described is, of course, just the well-known Hadley cell. From (11.38) we now see that in the absence of eddy transports of heat and momentum the circulation forced by solar differential heating will be the Hadley cell circulation. In the equatorial area where the eddies are very weak and the heating is strong the observed meridional circulation is indeed that of a direct Hadley cell. However, in the middle-latitude troposphere where both eddy heat flux and eddy momentum fluxes are strong, the eddy terms will tend to balance the diabatic heating term in (11.38) so that the actual meridional circulation is only a small residual. For this reason, the mean meridional motion is difficult to measure directly from zonal averaging of the observed data. However, if the eddy flux terms and the differential heating can be determined, it is possible to deduce the mean meridional motion by solving (11.37) just as the vertical motions in a baroclinic wave can be deduced indirectly from the omega equation. Once $\bar{v}$ has been deduced in this manner, (11.36) can be used to determine the steady state mean zonal wind distribution implied by the simple linear drag friction law. Present evidence, both from direct measurements of $\bar{v}$ and indirect deduction from (11.37), indicate that the mean meridional circulation is indirect in the middle latitude troposphere, but that it is so weak that it must play only a secondary role in the maintenance of the middle-latitude zonal mean circulation.

11.5 Laboratory Simulation of the General Circulation

In the previous sections we have seen that the gross features of the general circulation can be understood within the framework of the quasi-geostrophic model. In particular, we found that the observed zonal mean circulation and the atmospheric energy cycle are both rather well described by quasi-geostrophic theory. The fact that the quasi-geostrophic model, in which the spherical earth is replaced by the β plane, can successfully model many of the essential features of the general circulation suggests that the fundamental properties of the general circulation are not dependent on the spherical geometry or other parameters unique to the earth, but may be common to all rotating differentially heated fluids. That this conjecture is in fact correct can be demonstrated in the laboratory with rather simple apparatus.

Laboratory simulation experiments are a valuable method of testing hypotheses concerning the nature of the general circulation because parameters in experiments are subject to external control. For example, the rotation rate or the intensity of differential heating may be varied in an experiment, whereas in the atmosphere we can only accept the conditions given by nature.

In the past several years there have been a number of experimental studies directed toward an understanding of rotating differentially heated fluid systems. In most of these experiments the working fluid has been water rather than air, primarily because it is easier to follow motions by various sorts of tracers in a liquid than in a gas. In one group of experiments the apparatus consists of a cylindrical vessel which is rotated about its vertical axis. The vessel is heated at its rim and cooled at the center. These experiments are often referred to as "dishpan" experiments since some of the first experiments of this type utilized an ordinary dishpan. The fluid in the dishpan experiments crudely represents one hemisphere in the atmosphere with the rim of the dishpan corresponding to the equator and the center to the pole. However, because the geometry is cylindrical rather than spherical, the β effect is not modeled for baroclinic motions.[2] Thus the dishpan experiments omit the dynamical effect of the earth's vorticity gradient as well as the geometrical curvature terms which were neglected in the quasi-geostrophic model.

Qualitatively the experiments indicate that for certain combinations of rotation and heating rates, the flow in the cylinder appears to be axially symmetric with a steady zonal flow which is in thermal-wind equilibrium with the radial temperature gradient, and a superposed direct meridional circulation with rising motion near the rim and sinking near the center. This symmetric flow is usually called the *Hadley regime* since the flow is essentially that of a Hadley cell.

However, for other combinations of rotation and heating rates the observed flow is not symmetric but consists, rather, of irregular wavelike fluctuations and meandering zonal jets. In such experiments velocity tracers on the surface of the fluid reveal patterns very similar to those on middle-latitude upper-air weather charts. This type of flow is usually called the *Rossby regime*, although it should be noted that the observed waves are not Rossby waves since there is no β effect in the tank.

In addition to the dishpan experiments a number of similar experiments have been carried out in which the fluid is contained in the annular region

---

[2] For barotropic motions the radial height gradient of the rotating fluid in the cylinder creates an "equivalent" β effect (see Section 5.3).

between two coaxial cylinders of different radii. In these experiments the walls of the inner and outer cylinders are held at constant temperatures so that a precisely controlled temperature difference can be maintained across the annular region. In these "annulus" experiments very regular wave patterns can be obtained for certain combinations of rotation and heating. In some cases the wave patterns are steady, while in other cases they undergo regular periodic fluctuations called "vacillation" cycles (see Fig. 11.8). Although these flow patterns do not have the striking resemblance to atmospheric flow patterns that can be seen in the dishpan experiments, the annulus experiments have had a greater scientific impact than the dishpan experiments because the regularity of the waves in the annulus and the abrupt transitions between flow regimes as external parameters are changed make the annulus experiments far more amenable to theoretical analysis. For this reason, the annulus experiments have led to important advances in our understanding of thermal convection in rotating fluid systems.

In order to determine whether the laboratory experiments are really valid analogues of the atmospheric circulation or merely bear an accidental resemblance to atmospheric flow, it is necessary to analyze the experiments quantitatively. The mathematical analysis of the experiments can proceed from essentially the same equations which are applied to the atmosphere except that cylindrical geometry replaces spherical geometry and temperature replaces potential temperature in the heat equation. (Since water is nearly incompressible, *adiabatic* temperature changes are negligible following the motion.) In addition the equation of state must be replaced by an appropriate measure of the relationship between temperature and density:

$$\rho = \rho_0[1 - \varepsilon(T - T_0)] \qquad (11.39)$$

where ε is the thermal expansion coefficient ($\varepsilon \simeq 2 \times 10^{-4}$ °C^{-1} for water) and ρ_0 is the density at the mean temperature T_0.

As we saw already in Chapter 1, the character of the motion in a fluid is crucially dependent on certain characteristic scales of parameters such as the velocity, pressure fluctuations, length, time, etc. In the laboratory the scales of these parameters are generally many orders of magnitude different from their scales in nature. However, it is still possible to produce quasi-geostrophic motions in the laboratory provided that the motions are slow enough so that the flow is approximately hydrostatic and geostrophic. As we saw in Section 2.4.2, the geostrophy of a flow does not depend on the absolute value of the scaling parameters, but rather on a nondimensional ratio of these parameters called the Rossby number. For the annulus experiments, the maximum horizontal scale is set by the dimensions of the tank so that it is convenient to define the Rossby number as

$$Ro \equiv \frac{U}{\Omega(b - a)}$$

where Ω is the angular velocity of the tank, U is a typical relative velocity of the fluid, and $b - a$ is the difference between the radius b of the outer wall and the radius a of the inner wall of the annular region.

Using the hydrostatic approximation

$$\frac{\partial p}{\partial z} = -\rho g \tag{11.40}$$

Together with the geostrophic relationship

$$2\Omega \mathbf{V}_g = \frac{\mathbf{k} \times \nabla p}{\rho_0} \tag{11.41}$$

and the equation of state (11.39) we can obtain a "thermal wind" relationship in the form

$$\frac{\partial \mathbf{V}_g}{\partial z} = \frac{\varepsilon g}{2\Omega} \mathbf{k} \times \nabla T \tag{11.42}$$

Letting U denote the scale of the geostrophic velocity, H the mean depth of fluid, and ΔT the radial temperature difference across the width of the annulus, we obtain from (11.42)

$$U \sim \frac{\varepsilon g H \, \Delta T}{2\Omega(b - a)} \tag{11.43}$$

Substituting the value of U in (11.43) into the formula for the Rossby number yields the *thermal Rossby number*

$$Ro_T = \frac{\varepsilon g H \, \Delta T}{2\Omega^2 (b - a)^2} \tag{11.44}$$

This nondimensional number is the best measure of the range of validity of quasi-geostrophic dynamics in the annulus experiments. Provided $Ro_T \ll 1$ the quasi-geostrophic theory should be valid for motions in the annulus. For example, in a typical experiment

$$H \simeq b - a \simeq 10 \quad \text{cm}, \qquad \Delta T \simeq 10°\text{C}, \qquad \text{and} \qquad \Omega \simeq 1 \quad \text{sec}^{-1}$$

so that

$$Ro_T \simeq 10^{-1}$$

There is, however, one difficulty with the thermal Rossby number as derived here. The difficulty arises because near the vertical boundaries of the annulus there are *conduction boundary layers* in which the temperature

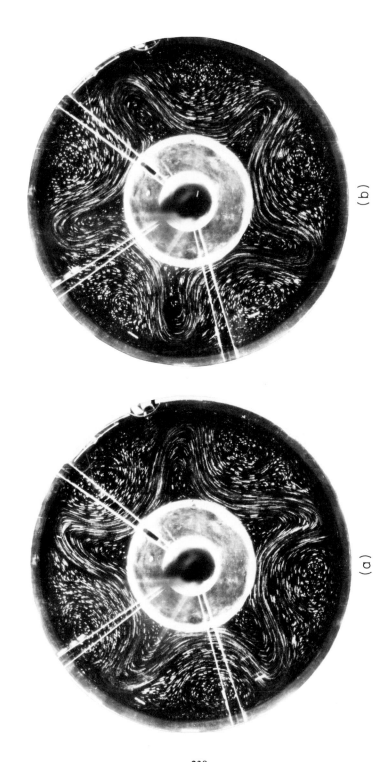

(b)

(a)

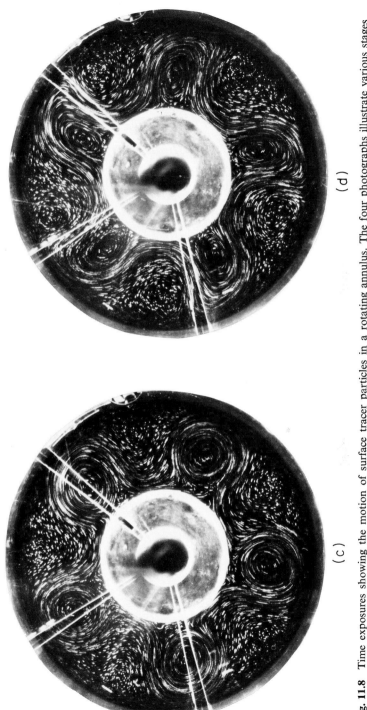

(c)

(d)

Fig. 11.8 Time exposures showing the motion of surface tracer particles in a rotating annulus. The four photographs illustrate various stages of a five-wave tilted trough vacillation cycle. The period of the vacillation cycle is 16¼ revolutions and the photographs are at intervals of four revolutions. (Photographs by Dave Fultz)

changes rather rapidly away from the boundaries. The thermal wind relationship is not valid in these boundary layers; thus the ΔT in (11.44) really
represents only the temperature gradient across the interior region of fluid
away from the sidewall boundary layers. However, this ΔT is not an external
parameter, but is in itself dependent on the flow. Therefore, in order to
describe the experimental results entirely in terms of external parameters, it is
usual to define an *imposed* thermal Rossby number

$$Ro_T{}^* \equiv \frac{\varepsilon g H}{2\Omega^2 (b-a)^2} (T_b - T_a) \qquad (11.45)$$

where T_b and T_a are the imposed temperatures of the outer and inner walls,
respectively, From the discussion above it is clear that $Ro_T{}^* > Ro_T$, so that
$Ro_T{}^* \ll 1$ is certainly a sufficient condition for quasi-geostrophic theory to be
valid, but is not clearly a necessary condition.

Annulus experiments carried out over a wide range of rotation rates and
temperature contrasts produce results which can be classified consistently on a
log-log plot of the imposed thermal Rossby number versus a nondimensional
measure of the rotation rate, $(G^*)^{-1} \equiv (b-a)\Omega^2/g$. The results for Fultz's
experiments are summarized in Fig. 11.9. The heavy solid line separates the
axially symmetric Hadley regime from the wavy Rossby regime. These results
can best be understood qualitatively by considering experiment in which
the thermal Rossby number is slowly increased from zero by gradually
imposing a temperature difference $T_b - T_a$. For very weak differential heating
($T_b - T_a$ very small) the motion is of the Hadley type with weak horizontal and
vertical temperature gradients in the fluid. In the absence of viscosity this flow
would probably be baroclinically unstable. Viscous boundary layer dissipation
removes the possibility of unstable wave growth. However, as the horizontal
temperature contrast is increased, the mean thermal wind must also increase
until at some critical value of $Ro_T{}^*$ the flow becomes baroclinically unstable.
According to the theory presented in Chapter 10, the wavelength of maximum
instability is proportional to the ratio of the static stability to the square of the
rotation rate. Thus, as can be seen in Fig. 11.9, the wave number observed
when the flow becomes unstable decreases (that is, wavelength increases) as
the rotation rate is reduced. Furthermore, since baroclinic waves transport
heat vertically as well as laterally, they will tend to increase the static stability
of the fluid. Therefore, as the thermal Rossby number is increased within
the Rossby regime the increased heat transport by the waves will raise the
static stability, and hence increase the wavelength of the wave of maximum
instability. Thus, as $Ro_T{}^*$ is increased the flow will undergo transitions in
which the observed wave number decreases until finally the static stability
becomes so large that the flow is stable to even the largest wave that can fit

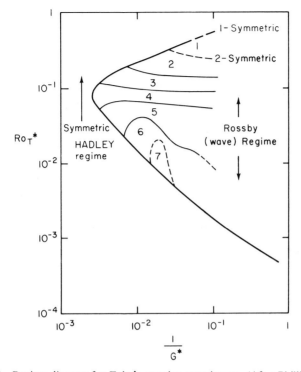

Fig. 11.9 Regime diagram for Fultz's annulus experiments. (After Phillips, 1963.)

the tank. The flow then returns to a symmetric Hadley circulation stabilized by a high static stability which is maintained by a vigorous direct meridional circulation. This regime is usually called the *upper* symmetric regime to distinguish it from the *lower* symmetric regime which occurs for very weak heating.

In the annulus where the flow is highly constrained by the geometry of the system the transitions between wave numbers in the Rossby regime as indicated in Fig. 11.9 are quite abrupt and regular, at least for moderately small values of $(G^*)^{-1}$. In the dishpan experiments, the waves are not so regular and transitions are more continuous. However, the basic features of the flow do remain the same.

So far we have discussed only the qualitative similarity between the Rossby regime and the general circulation of the atmosphere. Detailed temperature and velocity measurements have revealed that the energy cycle of the Rossby regime flow in the annulus and dishpan experiments is the same as that observed for the troposphere as shown in Fig. 11.1. Furthermore, the waves in the experiments under some conditions have an eddy momentum transport

$\langle u'v' \rangle$ similar to that observed in the atmosphere, with eddy momentum convergence in the region of strong zonal jets. In fact, it is this interchange of momentum between the eddies and the zonal flow which accounts for the vacillation cycle in the annulus shown in Fig. 11.8. It is clear from Fig. 11.8a that during this stage of the vacillation cycle the troughs and ridges have a strong southwest to northeast tilt (assuming that the axis of rotation corresponds to the North Pole) and that as a consequence there is a strong northward eddy momentum transport. As a result of this eddy momentum transport the zonal flow is concentrated in a meandering jet stream. Temperature measurements in this type of experiment have revealed that the horizontal temperature gradient is concentrated in a "frontal" zone associated with the meandering jet, as would be required for the jet to satisfy the thermal-wind equation.

In summary, the laboratory studies despite their many idealizations can model the most important of those features of the general circulation which are not dependent on the earth's topography or continent–ocean heating contrasts. Specifically, we find, perhaps surprisingly, that the β effect is not essential for the development of circulations which look very much like tropospheric synoptic systems. Apparently the role of the β effect is primarily a modifying influence which, although it certainly changes the zonal phase speed of atmospheric waves, is not essential to modeling the primary features of the middle-latitude circulation. Thus, the observed middle latitude waves should be regarded essentially as *baroclinic waves* modified by the β effect, not as Rossby waves in a baroclinic current.

In addition to demonstrating the primacy of baroclinic instability, the experiments also confirm that condensation heating is not an essential mechanism for simulation of midlatitude circulations. The laboratory experiments thus enable us to separate the essential mechanisms from second-order effects in a manner not easily accomplished by observation of the atmosphere itself. Also, since laboratory simulation experiments have typical rotation rates of 10 rpm it is possible to simulate many years of atmospheric flow in a short time. Thus, the model experiments provide a unique opportunity for collecting "general circulation" statistics. In addition, since temperatures and velocities can be measured at uniform intervals in the experiments, the experiments can provide excellent sets of data for testing long-term numerical weather prediction models.

11.6   The Numerical Simulation

In the previous section we have discussed the role of laboratory experiments in contributing toward a qualitative understanding of the general circulation of the atmosphere. Although laboratory experiments can elucidate most of the gross features of the general circulation, there are many details which cannot possibly be duplicated in the laboratory. For example, the possible long-term climatic effects of aerosols and trace gases which are added to the atmosphere as a result of man's activities could not possibly be predicted on the basis of laboratory experiments. As another example, the influence of an ice-free Arctic Ocean on the general circulation would also be extremely difficult to simulate in the laboratory. Since all the conditions of the atmosphere cannot be duplicated in the laboratory, it would appear that the only practical manner in which to predict possible climate modifications resulting from intentional or unintentional human intervention is by numerical simulation on high-speed computers.

Progress in numerical modeling of the general circulation has been to some degree dictated by the rate of development in the field of computer technology. However, our limited ability to parameterize the effects of small-scale processes in terms of the large-scale motions has been an equally important limiting factor. Essentially, the problem of numerical modeling of the general circulation is simply that of producing a very long-range numerical weather forecast. However, the equations used in general circulation models must be more sophisticated than those in numerical weather prediction models because as the length of the "forecast" is extended, physical processes which are unimportant for the short-term evolution may become crucial. For example, the interaction between the atmosphere and the upper layers of the ocean is probably important for understanding variations in the general circulation with periods greater than a month or so. On the other hand, since a forecast of the flow evolution from a specific initial state is not needed, the problems associated with specifying initial data in primitive equation forecast model do not arise in general circulation models. The object of most general circulation models is to simulate the climate. Since climate (here defined as the time-averaged circulation) should be independent of the initial conditions, it is customary to begin most general circulation model integrations with a resting atmosphere as the initial state.

Because the values of the field variables in a numerical model are known on a regular array of grid points at regular time intervals, it is very simple to compute the statistics needed to completely diagnose the energy cycle and momentum budget for a model general circulation. Thus, a general circulation model can be used like the dishpan experiments to test the quasi-geostrophic energy equations discussed in Section 11.2. In the first subsection below we will discuss a very simple quasi-geostrophic general circulation model suitable for this purpose. In the second subsection we will describe briefly some results from a highly sophisticated primitive equation general circulation model, and in the final subsection we will indicate how general circulation models can be used to give some insight into the ultimate limits of atmospheric predictability.

11.6.1 A QUASI-GEOSTROPHIC MODEL

The success of the quasi-geostrophic model in short-range weather prediction suggests that for simulating the gross features of the general circulation such a model might be adequate provided that diabatic heating and frictional dissipation were included in a suitable manner. In fact, we have already shown in Section 11.2 that the quasi-geostrophic model is capable of at least qualitatively describing the observed energy cycle and momentum budget in the midlatitude troposphere.

Phillips (1956) made the first attempt to model the general circulation numerically. His classic experiment employed a two-level quasi-geostrophic forecast model similar to the model discussed in Section 8.5. However, the model was modified so that it included friction, primarily in the form of an Ekman layer dissipation. Phillips realized that to represent accurately the diabatic heating would have required extensive radiative transfer computations as well as description of the water vapor budget. In order to avoid such complexities, he simply specified a net heating rate, based on observations, which was a linear function of latitude only and has a zero horizontal mean. The domain of integration was a zonal channel on the β plane with a latitudinal width of 10,000 km and an assumed zonal periodicity with 6000-km wavelength. The heating rates chosen by Phillips were based on estimates of the *net* diabatic heating rates necessary to balance the poleward heat transport at 45°N computed from observational data. Despite the severe limitations of the computer available, and the eventual breakdown of the integration due to a nonlinear computational instability, the computations resulted in a model circulation which in many respects resembled the observed circulation. Perhaps this does not seem surprising since we know that the quasi-geostrophic model can produce reasonable short-range forecasts, and scaling considerations indicate that the quasi-geostrophic model contains the essential physical mechanisms for the development of synoptic scale eddies. However, it should be kept in mind that Phillips' experiment was quite a bold departure from the previous applications of the quasi-geostrophic model to short-term forecasts. In the forecast applications, the model equations were integrated from an observed initial state already containing fully developed systems. In Phillips' model the atmosphere was started from an initial state of rest and the motion was allowed to develop as a result of the differential diabatic heating.

The motion field which initially developed was a pure Hadley cell circulation with rising motion south of the center of the channel and sinking motion north of the center. As the diabatic heating increased the latitudinal temperature gradient, the zonal thermal wind also increased until the flow became baroclinically unstable At that point, small random perturbations introduced into the calculations caused the development of baroclinic waves. These waves grew rapidly until the rate of energy conversion from the mean flow was approximately balanced by frictional dissipation in the waves. Although the accumulation of truncation errors in Phillips' experiment prevented the flow from reaching a statistically steady state, the circulation for about 20 model days after the initiation of baroclinic instability did remarkably resemble that of a midlatitude weather chart as Fig. 11.10 reproduced from Phillips' paper indicates.

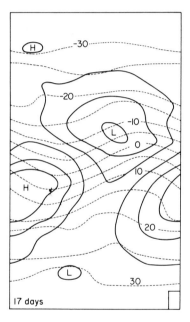

Fig. 11.10 Distribution of 1000-mb contour height at 200-ft intervals (solid lines) and 500-mb temperature at 5°C intervals (dashed lines) at 17 days. The small rectangle in the lower right corner shows the size of the finite-difference grid intervals Δx and Δy. (After Phillips, 1956.)

Phillips computed the energy transformations in his model and found that the energy cycle of the model atmosphere agreed remarkably with that shown in Fig. 11.1. Examination of the momentum budget indicated further that even the transfer of eddy momentum to mean zonal momentum was modeled in this simple experiment. Therefore, Phillips' numerical experiment provides additional evidence that the quasi-geostrophic model is capable of describing the essential characteristics of the general circulation outside the tropics.

11.6.2 PRIMITIVE EQUATION MODELS

Phillips' experiment, although it was an extremely important advance in dynamic meteorology, suffered from a number of shortcomings as a general circulation model. Perhaps the gravest shortcoming in his model was the specification of diabatic heating as a fixed function of latitude only. In reality the atmosphere must to some degree determine the distribution of its own heat sources. This is true not only for condensation heating, which obviously depends on the distribution of vertical motion and water vapor, but also holds for radiative heating as well. Because the long-wave radiative emission of the atmosphere is temperature dependent, warm eddies will tend to lose heat faster through long-wave radiation than will cold eddies. Thus radiation will tend to reduce the amplitude of the temperature perturbations, thereby acting as a sink for eddy available potential energy. This dissipative role of radiation was indicated by the $\{P' \cdot R'\}$ conversion term in Fig. 11.1.

Another important limitation of Phillips' two-level quasi-geostrophic model was that static stability could not be predicted (since temperature was computed at only a single level) but had to be specified as an external parameter. This limitation is a serious one because the static stability of the atmosphere is obviously controlled by the motions. Indeed, in the discussion of the annulus experiments in Section 11.5, we pointed out that the existence of the upper symmetric regime can be attributed to the stabilization of the flow due to the very high static stability established by the strong vertical heat transport of the meridional circulation. In the atmosphere the vertical temperature profile, and hence the static stability, is determined by the combined effects of diabatic heating and the vertical heat flux due to large-scale eddies and small-scale convection.

It would be possible to design a quasi-geostrophic model in which the diabatic heating and static stability were motion dependent. Indeed such models have been used to some extent, especially in theoretical studies of the annulus experiments. However, to model the global circulation clearly requires a dynamical framework which is valid in the equatorial zone. Thus, it is highly desirable to base a general circulation model on the primitive equations. Because, as was discussed in Chapter 8, gravity waves are not filtered by the primitive equations, time increments in a primitive equation model must be much smaller than for a quasi-geostrophic model. Therefore, an enormous amount of computation is necessary to simulate even a single season of the global circulation. However, the task is well within the capabilities of modern computers. Because of its enormous complexity and many important applications, general circulation modeling has become a highly specialized activity which cannot possibly be adequately covered in a short space. To keep the discussion within the scope of this text, we will here discuss only one of the primitive equation models, the nine-level moist model developed by J. Smagorinsky and his co-workers at the Geophysical Fluid Dynamics Laboratory of NOAA.

The Smagorinsky model is based on the primitive equations in the σ

coordinate system. The nine prediction levels are arranged at unequal vertical intervals designed to give an adequate resolution both in the planetary boundary layer near the ground and in the lower stratosphere. As initial conditions in the model, an isothermal atmosphere at 289°K is assumed. In the experiment described here, the solar radiation flux at the top of the atmosphere is specified to be the annual mean value. The radiative heating calculation includes the effects of ozone, carbon dioxide, and water vapor on the radiative heat balance. In addition the moist model has diabatic heating due to condensation of water vapor. Thus, the specific humidity must be carried as an independent variable in this model. The grid resolution of approximately 500 km horizontally is adequate to incorporate the condensation heating due to synoptic scale uplift. However, it is obviously not adequate to incorporate explicitly cumulus convection. The latter is implicitly included in the model through a *convective adjustment* process which operates as follows: Any time the lapse rate at some grid point exceeds moist adiabatic and the relative humidity is 100 percent, the lapse rate is adjusted back to the moist adiabatic value using a scheme which conserves the potential energy for the entire vertical column. This process at least crudely simulates the stabilizing effect of cumulus convection. In addition, any time that the lapse rate exceeds the dry adiabatic rate, it is similarly adjusted back to the dry adiabatic value.

Starting from the initial isothermal field, the model atmosphere develops its own vertical temperature profile as the integration proceeds. (The development of the temperature profile as a function of time is shown in Fig. 11.11.) At the same time the differential solar heating establishes a latitudinal temperature gradient. Just as in Phillips' model, the flow is initially a zonally symmetric Hadley circulation with the zonal wind in thermal wind balance with the latitudinal temperature gradient. After about 50 model days, this zonal flow becomes intense enough so that unstable baroclinic waves develop. By the 149th day of the integration, the temperature profile is very close to equilibrium. Therefore, flow statistics for this model can be computed using the period from 149 days to the end of the experiment at 187 days. A meridional section of the zonal mean temperature averaged over this period is shown in Fig. 11.12a. For comparison, the observed annual mean in the Northern Hemisphere is shown in part (b) of the figure. The model reproduces most of the features of the observed temperature structure quite well including the high cold tropical tropopause and the low polar tropopause. Thus, it is certain that this model contains the basic physical processes necessary for the formation and maintenance of the tropopause. The corresponding distribution of the mean zonal wind is shown in Fig. 11.13. The model produces a jet stream at approximately the right latitude; however, the speed of the jet is much greater than the observed annual mean. Actually, the computed jet is

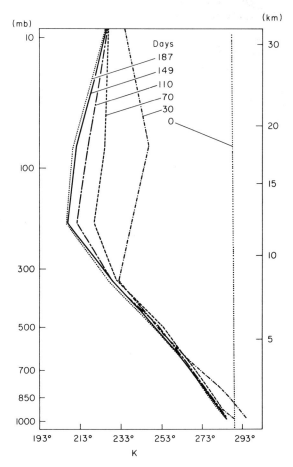

Fig. 11.11 Variation of the vertical distribution of temperature of the model atmosphere with time at 45° latitude. (After Manabe *et al.*, 1965.)

quite similar to the observed winter mean even though the annual mean solar heating was specified.

The energy cycle and momentum budget for this model circulation have been completely diagnosed. The energy cycle agrees quite well with the observed cycle indicated in Fig. 11.1. The poleward transport of angular momentum by various processes in the model is shown in Fig. 11.14. The large equatorward eddy transport in the tropics is unrealistic (it did not appear in a similar model which omitted the hydrological cycle). However, the general pattern agrees well with observations in that the eddy momentum flux is a maximum in the middle latitudes while the flux by the mean meridional circulation is a maximum in the tropics.

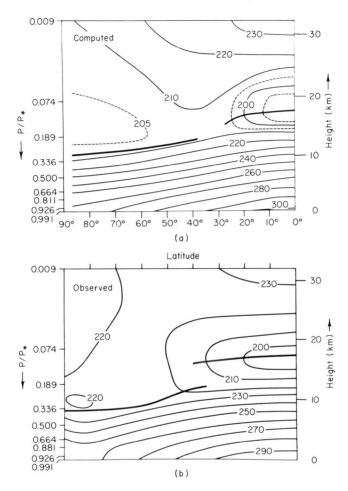

Fig. 11.12 Latitude–height distribution of temperature of the model atmosphere (a) and that of the actual atmosphere (b). (After Manabe *et al.* 1965.)

A comparison of the results from this model with those of a similar run without a hydrological cycle indicate that in middle latitudes there is little difference between the general circulation of a dry atmosphere and that of a wet atmosphere. However, in the tropics the circulation for a dry atmosphere is much weaker than that of the moist model. Thus, it appears that latent heat release is an essential driving force for the tropical atmosphere, but not for the extratropical regions. This comparison illustrates one of the most powerful features of the numerical models, namely that various physical processes can be included or excluded at will by modifying the equations; and by

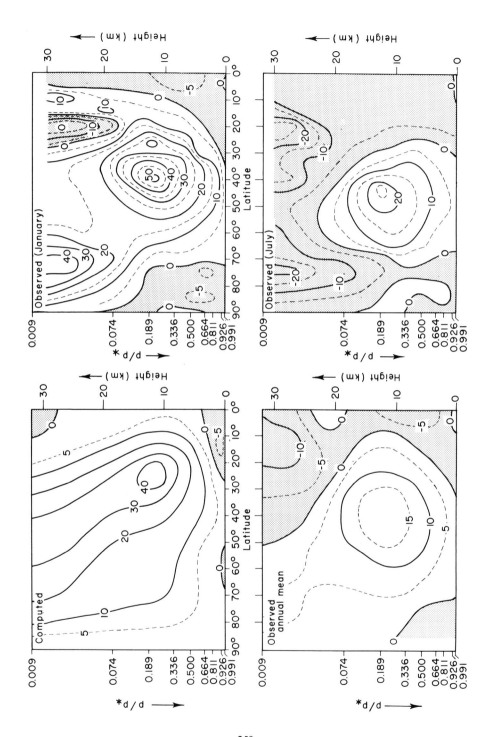

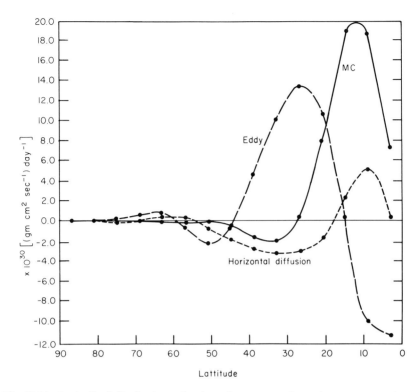

Fig. 11.14 Latitudinal distributions of poleward transport of angular momentum by the meridional circulation (MC), the large-scale eddies (Eddy), and the horizontal subgrid-scale mixing (Horizontal diffusion). (After Manabe *et al.*, 1965.)

comparing the output from runs with and without a given process, the effect of that process on the circulation can be easily evaluated. Using this sort of procedure, it should eventually be possible to make very good predictions of the climatological consequences of various man-caused environmental changes.

11.6.3 ATMOSPHERIC PREDICTABILITY

One of the most interesting possible applications of global general circulation models is to provide extended-range numerical forecasts. In Chapter 8

Fig. 11.13 Latitude–height distribution of the zonal mean of the zonal wind. The computed distribution and the observed annual mean are shown in the upper left and lower left, respectively. The distributions at 80°W for both January and July are shown on the right-hand side of the figure. (After Manabe *et al.*, 1965.)

we indicated that for short range forecasts (one or two days) of the midlatitude 500-mb flow, it is possible to neglect diabatic heating and frictional dissipation. However, it is important to have good initial data in the region of interest because on this short time scale, forecasting ability depends primarily on the proper advection of the initial vorticity field. As the length of the forecast period increases, the effects of various sources and sinks of energy become increasingly important. Therefore, the flow at a point in the middle-latitude troposphere will depend on initial conditions for an increasing domain as the period of the forecast is extended. In fact, according to the estimate of Smagorinsky shown in Fig. 11.15, for periods greater than a week, it is

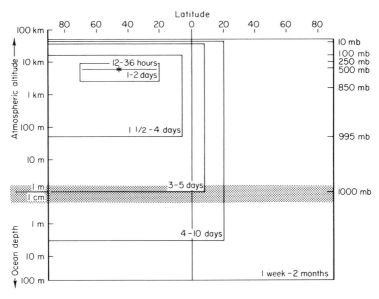

Fig. 11.15   A schematic diagram of the domain of initial data dependence for a prediction point in midtroposphere at midlatitudes (denoted by a star) as a function of forecast timespan. The atmospheric and ocean elevations are given on a logarithmic scale, increasing upward and downward, respectively. The stripped area is the interface zone. (After Smagorinsky, 1967.)

necessary to know the initial state of the entire global atmosphere from the stratosphere to the surface as well as the state of the upper layers of the oceans.

However, even if the ideal data network were available to specify the initial state on a global scale, there still would be a limit beyond which a useful forecast would not be possible. This limit to the predictability of the atmosphere arises because the atmosphere is a continuum with a continuous spectrum of scales of motion. No matter how fine the grid resolution is made, there will always be motions whose scales are too small to be properly represented in the model. Such motions will inevitably influence the initial data and be interpreted by the model as errors in the initial data. As the forecast proceeds these so-called *aliasing* errors will gradually cause the forecast flow field to evolve differently from the actual flow field. Estimates of how this effect limits the inherent predictability of the atmosphere have been made in several ways. Perhaps the most convincing estimates have been derived from the general circulation models.

Smagorinsky's group has used the nine-level, hemispheric, moist general circulation model in a group of experiments designed to study the predictability problem. In these experiments a control run was first carried out using a set of initial data corresponding to the observed hemispheric flow on a given day. The initial data set was then perturbed by introducing a random temperature error of 0.5°C standard deviation. The model was then run again and the second experiment regarded as a "forecast" of the control run. The accuracy of the forecast measured by the average temperature error in the forecast is plotted versus time in Fig. 11.16. In order to indicate the skill of the forecast

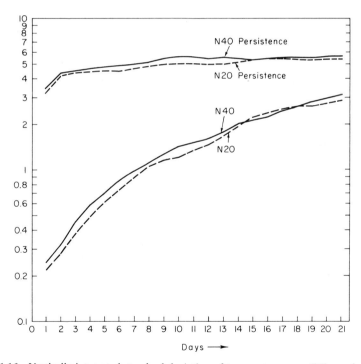

Fig. 11.16 Vertically integrated standard deviation of temperature error (°C) as a function of forecast time interval for two different computational resolutions with 20 and 40 grid points, respectively, from equator to pole. The upper two curves are for a persistence forecast and the lower two for the difference between the control and the initially perturbed temperature field. (After Smagorinsky, 1969.)

the average temperature error for a forecast of persistence (that is, no change from the initial state) is also shown. Notice that initially the temperature error actually decreases. This is due to adjustment of the perturbed temperature field to a quasi-geostrophic balance with the wind field. However, after the initial adjustment the temperature errors grow with a doubling time which is about two days initially, but gradually increases as the forecast progresses. Thus, the evidence from these experiments is that the theoretical limit of predictability on the synoptic scale may be at least three weeks. However, with the present data network and present models the practical limit of predictability is far shorter.

11.7   The General Circulation of the Stratosphere

So far in this chapter we have focused almost exclusively on the troposphere. The troposphere contains most of the mass and energy in the atmosphere and it is primarily the energy transformations in the troposphere which produce our weather. However, the circulation of the stratosphere is of interest both for its own sake and because we probably must include the interaction of the troposphere and stratosphere in very long-range forecast models. In the following subsections some of the most interesting aspects of the stratospheric circulation will be briefly considered.

11.7.1   THE ENERGY CYCLE OF THE LOWER STRATOSPHERE

In the lower stratosphere baroclinic instability is suppressed by the extremely high static stability. The energy cycle in the lower stratosphere must, therefore, be rather different from the energy cycle in the troposphere. It is now fairly certain both from observational and theoretical evidence that the eddy energy of the lower stratosphere is provided by vertical propagation of kinetic energy from the troposphere. Theory suggests that only planetary waves of very long wavelength can propagate significant amounts of energy vertically, and this is indeed consistent with the observed predominance of long waves in the stratosphere. As was indicated in the temperature profile in Fig. 11.12, the polar stratosphere is warmer than the equatorial stratosphere in the annual mean. Furthermore, there is evidence that radiative processes tend to heat the equatorial stratosphere and cool the polar stratosphere in the mean so that the observed positive poleward temperature gradient cannot be accounted for on the basis of radiative heating. Observations in the Northern Hemisphere have now confirmed that the stratospheric energy cycle is indeed reversed from that of the troposphere. The eddies in the stratosphere transport heat northward just as the eddies in the troposphere. But because the mean temperature gradient is reversed from that in the troposphere, the eddy heat transport in the stratosphere is *up the temperature gradient*. The stratosphere then (in the mean) is thermodynamically equivalent to a refrigerator. The eddies remove heat from the cold equatorial region and deposit it in the warm polar region thereby maintaining the reversed mean temperature gradient against the radiative damping. This "refrigerator" is driven by the tropospheric heat engine through the vertical propagation of kinetic energy. In Fig. 11.17 this energy cycle is shown schematically in a block diagram

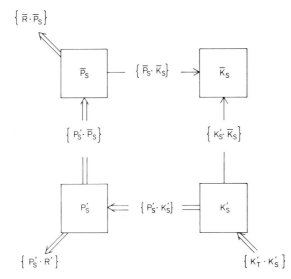

Fig. 11.17 The energy cycle in the lower stratosphere. Subscript S indicates stratosphere, subscript T indicates troposphere. Thus, $\{K_T' \cdot K_S'\}$ is the conversion of tropospheric kinetic energy to stratospheric kinetic energy.

similar to that of Fig. 11.1 for the troposphere. The primary transformations are from eddy kinetic energy to eddy potential energy due to the sinking of warm air relative to cold air in the eddies and the simultaneous transformation of eddy available potential energy to zonal mean available potential energy by the northward advection of warm air and southward advection of cold air in the eddies. Referring back to the arguments given in Section 11.4, we see that the northward eddy heat flux in the stratosphere will tend to create a direct mean meridional circulation with rising motion in the warm polar stratosphere and subsidence in the cold equatorial stratosphere. As a result of this direct meridional circulation, some of the zonal available potential enegy is converted to zonal mean kinetic energy. Radiation provides an additional sink for the zonal available potential energy.

Thus, from an energetic standpoint the stratosphere in the annual mean acts as a rather passive layer which is primarily driven by leakage of energy from the lower layers. The stratosphere, however, does have many types of motion of great interest which occur as a direct result of this energy leakage from the troposphere. The two phenomena of this type which have received the most attention in recent years are the *sudden warming* in the winter polar stratosphere and the *quasi-biennial oscillation* in the equatorial stratosphere.

11.7.2   THE SUDDEN WARMINGS

Although the latitudinal temperature gradient in the annual mean is positive from equator to pole in the stratosphere the situation for a winter seasonal mean is quite different. In the normal winter situation the temperature in the lower stratosphere reaches a maximum at about 45° latitude and decreases poleward so that the polar and equatorial stratospheric temperatures are almost equally cold. From thermal wind considerations this cold polar stratosphere requires a zonal vortex with strong westerly shear with height.

Every few years this normal winter pattern of a cold polar stratosphere with a westerly vortex is interrupted in a spectacular manner in midwinter. Within the space of a few days the polar vortex becomes highly distorted and breaks down with an accompanying large-scale warming of the polar stratosphere which can quickly reverse the meridional temperature gradient and create a circumpolar easterly current. In some cases warmings of as much as 40°C in a few days have occurred at the 50-mb level as shown in Fig. 11.18. Numerous studies of the energetics of the sudden warming confirm that enhanced propagation of energy from the troposphere by planetary scale waves, primarily zonal wave numbers 1 and 2, is essential for the development of the warming. Since the sudden warming is observed only in the Northern Hemisphere, it is logical to conclude that topographically forced waves are responsible for the vertical energy propagation. The Southern Hemisphere with its relatively small land masses at middle latitudes has much smaller amplitude stationary planetary waves. However, even in the Northern Hemisphere, it is apparently only in certain winters that conditions are right to allow sufficient vertical energy propagation to produce the sudden warming.

11.7.3   THE QUASI-BIENNIAL OSCILLATION

The search for periodicities in atmospheric motions has a long history. However, aside from the externally forced diurnal and annual components and their harmonics, no evidence of truly periodic oscillations has been found. Probably the phenomenon which comes closest to exhibiting periodic behavior is the quasi-biennial oscillation in the mean zonal winds of the equatorial stratosphere. This oscillation has the following observed features: Zonally

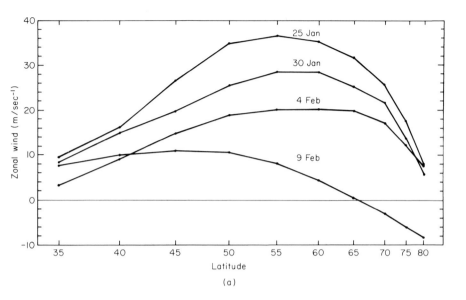

(a)

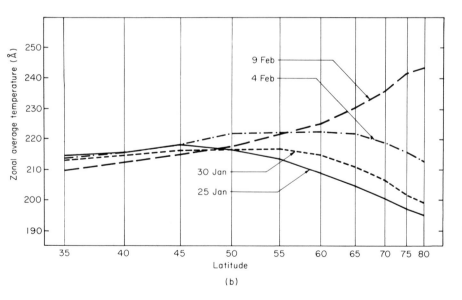

(b)

Fig. 11.18 Variation with latitude and time at the 50-mb level of (a) the zonal wind, and (b) the zonal mean temperature during the sudden warming of 1957. (After Reed *et al.*, 1963).

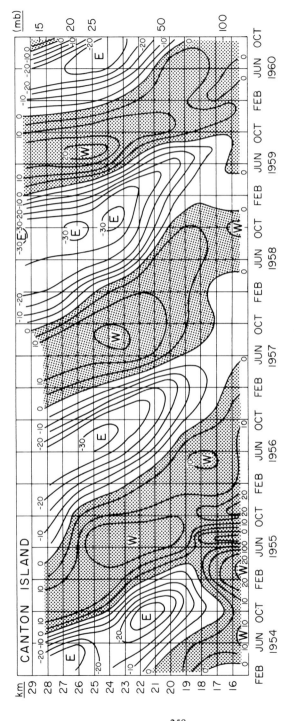

Fig. 11.19 Time-height section for Canton Island (2°46'S, 171°43'W), February 1954–October 1960. Isopleths are monthly mean zonal wind components in meters per second. Negative values denote easterly winds. (After Reed and Rogers, 1962.)

symmetric easterly and westerly wind regimes alternate regularly with a period varying from about 24 to 30 months. Successive regimes first appear above 30 km, but propagate downward at a rate of 1 km month^{-1}. The downward propagation occurs without loss of amplitude between 30 and 23 km, but there is rapid attenuation below 23 km. The oscillation is symmetric about the equator with a maximum amplitude of about 20 m sec^{-1}, and a half-width of about 12° latitude. This oscillation is best depicted by means of a time–height section of the zonal wind speed at an equatorial station as shown in Fig. 11.19. It is apparent from the figure that the vertical shear of the wind is quite strong at the level where one regime is replacing the other. Because this oscillation is zonally symmetric and symmetric about the equator, the zonal wind is in geostrophic balance nearly all the way to the equator. Thus, there must be a relatively strong meridional temperature gradient in the vertical shear zone to satisfy the thermal wind balance.

A complete theoretical model of the quasi-biennial oscillation is beyond the scope of this book. The main factors which a theoretical model of the quasi-biennial oscillation must explain are the approximate biennial periodicity, the downward propagation without loss of amplitude, and the occurrence of zonally symmetric westerlies at the equator. Since a zonal ring of air in westerly motion at the equator has an angular momentum greater than that of the earth, no symmetric advection process could account for the westerly phase of the oscillation. Therefore, there must be an eddy momentum source to produce the westerly accelerations in the downward propagating westerly shear zone. Observational studies by Wallace and Kousky (1968) and the theoretical model of Lindzen and Holton (1968) have confirmed that planetary scale vertically propagating internal gravity waves are the momentum sources for the oscillation. According to the discussion of Section 9.4.3 (see especially Fig. 9.3), eastward moving gravity waves should carry westerly momentum upward while westward moving gravity waves should carry easterly momentum upward. If the westerly momentum carried by these waves is absorbed in westerly shear zones and the easterly momentum in easterly shear zones, the result in each case will be to propagate the shear zone downward. Thus, as was the case with the sudden warmings, the quasi-biennial oscillation owes its existence to the vertical propagation of planetary scale waves into the stratosphere.

Problems

1. Verify Eq. (11.13) by carrying out the steps in the derivation omitted in the text.

2. Starting with Eqs. (11.18) and (11.20) derive the mean and eddy kinetic energy equations (11.21) and (11.22).

3. Starting with Eqs. (11.23) and (11.24) derive the mean and eddy available potential energy equations (11.25) and (11.26).

4. Using the observed data given in Fig. 11.1, compute the time required for each possible energy transformation or loss to restore or deplete the observed energy stores. (a watt equals 1 joule sec^{-1})

5. Derive the expression (11.42) for the "thermal wind" in the dishpan experiments.

6. Compute an estimate of the amplitude of the temperature oscillation associated with the quasi-biennial oscillation assuming a vertical shear of the zonal wind of 20 m sec^{-1} at the equator and a half-width of $12°$ latitude.

7. Show that for the case $\bar{R} = d = 0$, the zonal mean potential vorticity equation (11.35) can be derived directly from the potential vorticity equation (8.14).

Suggested References

Lorenz, *The Nature and Theory of the General Circulation*, is an excellent introduction to the subject which has an extensive discussion of the observational aspects of the general circulation, a historical review of former theories of the general circulation, and a discussion of the current status of general circulation theory. The book also contains a useful bibliography of original papers on the subject.

Newell, (ed.), *Observational Aspects of the Tropical General Circulation*, gives a more complete and up to date account of the general circulation in the tropical areas than is contained in Lorenz's book. See especially the chapter by Holton and Wallace, "On the Nature of Large Scale Eddies in the Tropical Stratosphere."

Fultz (1961) reviews a number of types of laboratory simulation experiments for large-scale geophysical phenomena including the rotating annulus experiments. His paper also contains an extensive bibliography.

Manabe *et al.* (1965), give a complete description and diagnosis of the simulated climatology produced by their numerical model of the general circulation.

Chapter

12 | Tropical Motion Systems

Throughout the previous chapters of this book, we have emphasized the circulation systems of the extratropical regions (that is, the regions poleward of about 30° latitude). This emphasis has not been due to any lack of interesting motion systems in the tropics, but is a result, rather, of the relatively primitive state of knowledge concerning the dynamics of tropical circulations. Most meteorologists live in the middle latitudes of the Northern Hemisphere, and the best data coverage is also in that region. For these reasons, there has historically been a much greater effort devoted to the dynamics of extratropical systems. These efforts have culminated in the development of the quasi-geostrophic theory, which, as we have tried to emphasize in this book, provides a basic theoretical framework for understanding the dynamics of middle-latitude synoptic systems and their role in the general circulation of the atmosphere. It is primarily due to the existence of a reasonably satisfactory theory that middle-latitude synoptic systems have been emphasized in this book.

However, it is in the tropics that the majority of the solar energy which drives the atmospheric heat engine is absorbed by the earth and transferred to the atmosphere. Therefore, an understanding of the general circulation of the tropics must be regarded as a fundamental goal of dynamic meteorology. Furthermore, since the tropics cover half the surface of the earth, development of forecasting models for the tropics is an important goal in itself. In

261

addition, as mentioned in the previous chapter, the tropical and middle-latitude atmosphere are coupled so that for long-term numerical forecasts of the motions at middle latitudes accurate forecasts in the tropics are also required.

Unfortunately, there is as yet no single unifying theory for tropical motions comparable to the quasi-geostrophic theory for middle-latitude motions. Rapid progress is being made in the understanding of tropical motion systems. However, the problems involved are in some ways much more difficult than in middle latitudes. We have seen that outside the tropics the primary energy source for synoptic disturbances is the zonal available potential energy associated with the strong latitudinal temperature gradient. In fact, observations indicate that diabatic heating due to latent heat release and radiative heating is as a rule only a secondary energy source in midlatitude systems. In the tropics, on the other hand, the storage of available potential energy is very small due to the very small temperature gradients. Latent heat release appears to be the primary energy source, at least for those disturbances which originate within the equatorial zone. Most latent heat release in the tropics occurs in convective cloud systems. These convective clouds are generally embedded in large-scale circulations. But the convective elements themselves are inherently mesoscale circulations. Thus, there is a strong interaction between the mesoscale convection and large-scale circulations which is of primary importance for understanding tropical motion systems.

Of course, the tropical atmosphere and middle-latitude atmosphere cannot be treated in complete isolation from each other. In the subtropical regions ($\sim 30°$ latitude) circulation systems characteristic of both tropical and extratropical regions may be observed depending on the season and geographical location. To keep the discussion as simple as possible we will, therefore, in this chapter focus primarily on the zone well equatorward of $30°$ where the influence of middle-latitude systems should be a minimum.

12.1   Scale Analysis of Tropical Motions

Despite the uncertainties involved with the interaction between the convective and synoptic scales, some information on the character of synoptic scale motions in the tropics can be obtained through the methods of scale analysis. The scaling arguments can be carried out most conveniently if the governing equations are written in the log-pressure coordinate system. Referring to Section 9.4.3 we recall that the momentum equation, hydrostatic equation, continuity equation, and thermodynamic energy equation may be written as follows in the log-pressure system:

$$\frac{d\mathbf{V}}{dt} + f\mathbf{k} \times \mathbf{V} = -\nabla\Phi \tag{12.1}$$

$$\frac{\partial\Phi}{\partial z^*} = \frac{RT}{H} \tag{12.2}$$

$$\frac{\partial u}{\partial x} + \frac{\partial v}{\partial y} + \frac{\partial w^*}{\partial z^*} - \frac{w^*}{H} = 0 \tag{12.3}$$

$$\left(\frac{\partial}{\partial t} + \mathbf{V} \cdot \nabla\right)T + w^*\Gamma = \frac{1}{c_p}\dot{H} \tag{12.4}$$

Here $\dot{H}$ is the diabatic heating rate per unit mass and

$$\Gamma = \frac{\partial T}{\partial z^*} + \frac{RT}{c_p H} = \frac{T}{\theta}\frac{\partial\theta}{\partial z^*}$$

is the static stability parameter. For the tropical troposphere, $\Gamma \simeq 3°C \text{ km}^{-1}$ is nearly constant.

We now compare the magnitudes of the various terms in (12.1)–(12.4) for synoptic scale motions in the tropics. We define characteristic scales for the various field variables as follows:

$U \sim 10^3 \text{ cm sec}^{-1}$	Horizontal velocity scale
W	Vertical velocity scale
$L \sim 10^8 \text{ cm}$	Horizontal length scale
D	Depth scale
$\Delta\phi$	Geopotential fluctuation scale
$L/U \sim 10^5 \text{ sec}$	Time scale for advection

We have here assumed magnitudes for the horizontal length and velocity scales which are typical for observed values in synoptic systems both in the tropics and at middle latitudes. We now wish to show how the corresponding characteristic scales for vertical velocity and geopotential fluctuations are limited by the dynamic constraints imposed by continuity, momentum balance, and heat energy balance.

We first note that an upper limit on W is imposed by the continuity equation (12.3). Thus, following the discussion of Section 5.5,

$$\frac{\partial u}{\partial x} + \frac{\partial v}{\partial y} \lesssim \frac{U}{L}$$

But for motions with a vertical scale less than or equal to the scale height ($D \lesssim H$),

$$\frac{\partial w^*}{\partial z^*} - \frac{w^*}{H} \sim \frac{W}{D}$$

so that the vertical velocity scale must satisfy

$$W \lesssim \frac{D}{L} U \qquad (12.5)$$

We can estimate the magnitude of the geopotential fluctuation by scaling the terms in the momentum equation. For this purpose, it is convenient to compare the magnitude of the horizontal inertial acceleration,

$$\mathbf{V} \cdot \nabla \mathbf{V} \sim \frac{U^2}{L}$$

with each of the other terms in (12.1) as follows:

$$\frac{\partial \mathbf{V}/\partial t}{(\mathbf{V} \cdot \nabla \mathbf{V})} \sim 1 \qquad (12.6)$$

$$w^* \frac{\partial \mathbf{V}/\partial z^*}{(\mathbf{V} \cdot \nabla \mathbf{V})} \sim \frac{WL}{UD} \lesssim 1 \qquad (12.7)$$

$$\frac{f\mathbf{k} \times \mathbf{V}}{(\mathbf{V} \cdot \nabla \mathbf{V})} \sim \frac{fL}{U} = Ro^{-1} \qquad (12.8)$$

$$\frac{\nabla \Phi}{(\mathbf{V} \cdot \nabla \mathbf{V})} \sim \frac{\Delta \Phi}{U^2} \qquad (12.9)$$

We have previously shown that in middle latitudes where $f \sim 10^{-4}$, the Rossby number Ro is small so that to a first approximation the Coriolis force and pressure gradient force terms balance. In that case $\Delta \Phi \sim fUL$. In the equatorial region, however, $f \lesssim 10^{-5}$ and the Rossby number is of order unity or greater. Therefore, it is not appropriate to assume that the Coriolis force term balances the pressure gradient. In fact from (12.6)–(12.9) we see that in general the pressure gradient must be balanced by the inertial acceleration so that $\Delta \Phi \sim U^2 \sim 100 \text{ m}^2 \text{ sec}^{-2}$. Thus, the geopotential perturbations associated with equatorial synoptic scale disturbances will be an order of magnitude smaller than those for midlatitude systems of similar scale.

This constraint on the geopotential fluctuations in the tropics has profound consequences for the structure of synoptic scale tropical motion systems. These consequences can be easily understood by applying scaling arguments to the thermodynamic energy equation. It is first necessary to obtain an estimate of the temperature fluctuations. From the hydrostatic approximation (12.2) we have

$$T = \frac{H}{R} \frac{\partial \Phi}{\partial z^*} \sim \frac{H}{D} \frac{\Delta \Phi}{R} \qquad (12.10)$$

Thus, for systems whose vertical scale is comparable to the scale height,

$$T \sim \frac{U^2}{R} \sim 0.3 \quad °C$$

Therefore, deep tropical systems are characterized by practically negligible temperature fluctuations. Referring to the thermodynamic energy equation, we find that for such systems

$$\left(\frac{\partial}{\partial t} + \mathbf{V} \cdot \mathbf{V}\right) T \sim 0.3 \quad °C \, day^{-1}$$

In the absence of precipitation the diabatic heating is primarily due to long-wave radiation which tends to cool the troposphere at a rate of $\dot{H}/c_p \sim -1°C \, day^{-1}$. Since the actual temperature fluctuations are small, this radiative cooling must be approximately balanced by adiabatic warming due to subsidence. Thus, to a first approximation (12.4) becomes

$$w^* \Gamma \simeq \frac{\dot{H}}{c_p} \tag{12.11}$$

For the tropical troposphere, $\Gamma \sim 3°C \, km^{-1}$ and the vertical motion scale must satisfy

$$W \sim \frac{\dot{H}}{\Gamma c_p} \sim 0.3 \quad cm \, sec^{-1} \tag{12.12}$$

Therefore, in the absence of precipitation the vertical motion is constrained to be even smaller than in extratropical synoptic systems of a similar scale. Therefore, to a rather high degree of accuracy the horizontal motion field is governed by the vorticity equation,

$$\left(\frac{\partial}{\partial t} + \mathbf{V}_\psi \cdot \mathbf{V}\right)(\zeta + f) + (\zeta + f)\mathbf{V} \cdot \mathbf{V} = 0 \tag{12.13}$$

which may be obtained by taking the vertical component of the curl of (12.1), neglecting terms involving w^* and approximating $\mathbf{V}$ by $\mathbf{V}_\psi$ in the advection term.

For the midlatitude case, we have seen that since $\zeta \ll f$ the divergence term can be approximated as simply $f\mathbf{V} \cdot \mathbf{V}$. In the tropics $\zeta \sim f$ so that the relative vorticity cannot be neglected compared to f in the divergence term. However, the terms

$$\left(\frac{\partial}{\partial t} + \mathbf{V}_\psi \cdot \mathbf{V}\right)\zeta \sim \frac{U^2}{L^2} \sim 10^{-10} \quad sec^{-1}$$

and

$$\mathbf{V}_\psi \cdot \nabla f \sim U\beta \sim 10^{-10} \; \sec^{-1}$$

have the same magnitudes in the tropics as in middle latitudes. But since in the tropics $(\zeta + f) \sim 10^{-5} \; \sec^{-1}$ and for deep systems $\nabla \cdot \mathbf{V} \sim W/H \sim 0.3 \times 10^{-6} \; \sec^{-1}$ we see that outside precipitation zones

$$(\zeta + f)\nabla \cdot \mathbf{V} \sim 10^{-11} \; \sec^{-1}$$

so that to a first approximation (12.13) becomes simply

$$\left(\frac{\partial}{\partial t} + \mathbf{V}_\psi \cdot \nabla\right)(\zeta + f) = 0 \tag{12.14}$$

Thus, in the absence of condensation heating, tropical motions in which the vertical scale is comparable to the scale height of the atmosphere must be *barotropic*. Such disturbances cannot convert potential energy to kinetic energy. They must, therefore, be driven by lateral coupling either to mid-latitude systems or to precipitating tropical disturbances.

For precipitating synoptic systems in the tropics, the above scaling considerations require considerable modification. Precipitation rates in such systems are typically in the range of 2 cm day^{-1}. This implies condensation of $m_w = 2$ gm water for an atmospheric column of 1 cm^2 cross section. For a latent heat of condensation $L_c \simeq 2.5 \times 10^{10}$ ergs gm^{-1}, this precipitation rate implies an addition of heat energy to the atmospheric column of

$$m_w L_c \sim 5 \times 10^{10} \; \text{erg cm}^{-2} \, \text{day}^{-1}$$

If this heat is distributed uniformly over the entire atmospheric column of mass $p_0/g \simeq 1 \; \text{kgm cm}^{-2}$, then the average heating rate per unit mass of air is

$$\frac{\dot{H}}{c_p} \simeq \frac{L_c m_w}{c_p (p_0/g)} \sim 5°\text{C day}^{-1}$$

In reality the condensation heating due to deep convective clouds is not distributed evenly over the entire vertical column, but is a maximum between 300 and 400 mb where the heating rate may be $\sim 10°\text{C day}^{-1}$. In this case the approximate thermodynamic energy equation (12.11) implies that the vertical motion on the synoptic scale in precipitating systems must have a magnitude of order $W \sim 3 \; \text{cm sec}^{-1}$ in order that the adiabatic cooling can balance the condensation heating in the 300–400-mb layer.

Therefore, the average vertical motion in precipitating disturbances in the tropics is an order of magnitude larger than the vertical motion outside the disturbances. As a result the flow in these disturbances has a relatively large divergent component (at least in the upper troposphere where $\partial w^*/\partial z^*$ is large) and the dynamics cannot be accurately described by the barotropic vorticity equation. In fact, scaling arguments indicate that all terms of the complete vorticity equation should be retained to describe quantitatively the flow field. In addition there is some evidence that the combined effect of vertical transport of vorticity by the individual cumulus towers must be included in the vorticity budget for the synoptic scale. This, of course, introduces a complicated problem of interaction between the mesoscale convection and the synoptic scale. On the other hand, temperature fluctuations are observed to remain small in precipitating tropical systems (at least above the planetary boundary layer). Thus, the thermodynamic energy equation remains in the simple form (12.11) so that the synoptic scale vertical motion, and hence the divergence field, can be diagnostically computed if the synoptic scale distribution of diabatic heating is known. This matter will be discussed further in Section 12.2.5.

All the above scaling arguments have been based on the assumption that the depth scale of tropical disturbances is comparable to the scale height of the atmosphere. There is, however, one important class of large-scale tropical motions for which this assumption does not apply. These are the vertically propagating planetary scale gravity waves which were mentioned briefly in Section 11.7.3. Such vertically propagating waves have vertical scales which are generally small compared to the scale height. Hence, since

$$\mathbf{V} \cdot \mathbf{V} \sim \frac{W}{D}$$

these waves tend to have large horizontal divergence fields even though the vertical velocities remain quite small. Such waves are important for transporting momentum and energy into the stratosphere, but probably have little influence on tropospheric weather systems.

12.2 Cumulus Convection

The scaling arguments of the previous section suggest that to a large extent the structure of synoptic disturbances in the tropics must be determined by the field of heating provided by latent heat release in areas of active precipitation. Observations indicate that most precipitation in tropical systems is of convective origin and is concentrated in a comparatively small number of very deep vigorous cumulus convection cells. These so-called "hot towers" occupy only a small fraction of the area of the tropics even in active disturbances. Therefore, to understand the dynamics of synoptic scale systems in the tropics, it is necessary to elucidate the nature of the interaction between the cumulus scale motions and the synoptic scale disturbances in which the cumulus motions are embedded. This in turn requires an understanding of the dynamics of cumulus scale motions themselves.

The subject of convective motions is extremely complex to treat theoretically. Part of this difficulty results from the fact that convective motions are generally nonhydrostatic, nonsteady turbulent motions which require the full three-dimensional equations of motion for their description. In this section we, therefore, concentrate primarily on the *thermodynamic* aspects of moist convection.

12.2.1 EQUIVALENT POTENTIAL TEMPERATURE

We have previously applied the parcel method to discuss the vertical stability of a dry atmosphere. We found that the stability of a dry parcel with respect to a vertical displacement depends on the lapse rate of potential temperature in the environment such that a parcel displacement is neutrally stable provided that $\partial\theta/\partial z = 0$. The same condition also applies to parcels in a moist atmosphere when the relative humidity is less than 100 percent. However, if a parcel of moist air is forced to rise, it will eventually become saturated and a further rise will cause condensation and latent heat release. If the environmental lapse rate is greater than the moist adiabatic lapse rate, the parcel may then become buoyant relative to its surroundings and be accelerated upward. The criterion for this type of parcel instability, which is called *conditional instability*, can be expressed conveniently in terms of a quantity called *equivalent potential temperature*. Equivalent potential temperature, designated by θ_e, is the potential temperature that a parcel of air would have if all its moisture were condensed out and the resultant latent heat used to warm the parcel. The temperature of an air parcel can be brought to its equivalent potential value by raising the parcel from its original level until all the water vapor in the parcel has condensed and fallen out, then compressing the parcel adiabatically to a pressure of 1000 mb. Since the condensed water is assumed to fall out, the temperature increase during the compression will be at the dry adiabatic rate and the parcel will arrive back at its original level with a temperature which is higher than its original temperature. The process is, therefore, irreversible. Ascent of this type, in which all condensation products are assumed to fall out is called *pseudo-adiabatic ascent*. (It is not a truly adiabatic process because the liquid water which falls out carries a small amount of heat with it.)

A complete derivation of the mathematical expression relating θ_e to the other variables of state is rather involved and will be relegated to Appendix D. However, for most purposes, it is sufficient to use an approximate expression for θ_e which can be immediately derived from the entropy form of the first law of thermodynamics (1.20). If we let q_s denote the mass of water vapor per unit mass of dry air in a saturated parcel (q_s is called the saturation *mixing ratio*), then the rate of diabatic heating per unit mass is

$$\dot{H} = -L_c \frac{dq_s}{dt}$$

where L_c is the latent heat of condensation. Thus, from the first law of thermo-dynamics

$$c_p \frac{d \ln \theta}{dt} = -\frac{L_c}{T} \frac{dq_s}{dt} \tag{12.15}$$

For a saturated parcel undergoing pseudo-adiabatic ascent the rate of change in q_s following the motion is much larger than the rate of change in T or L_c. Therefore,

$$d \ln \theta \simeq -d\left(\frac{L_c q_s}{c_p T}\right) \tag{12.16}$$

Integrating (12.16) from the initial state (θ, q_s) to a state where $q_s \simeq 0$ we obtain

$$\ln \left(\frac{\theta}{\theta_e}\right) \simeq -\frac{L_c q_s}{c_p T} \tag{12.17}$$

where θ_e, the potential temperature in the final state, is approximately the equivalent potential temperature defined above. Thus, θ_e for a saturated parcel is given by

$$\theta_e \simeq \theta \exp\left(\frac{L_c q_s}{c_p T}\right) \tag{12.18}$$

The expression (12.18) may also be used to compute θ_e for an unsaturated parcel provided that the temperature used in the formula is the temperature which the parcel would have if expanded adiabatically to saturation. Thus, equivalent potential temperature is conserved for a parcel during both dry adiabatic and pseudo-adiabatic changes of state.

12.2.2 THE PARCEL METHOD

To derive the stability condition for a saturated parcel, we consider an environment in which the potential temperature is θ_0 at the level z_0. At the level $z_0 - \delta z'$ the undisturbed environmental air thus has potential temperature

$$\theta_0 - \frac{\partial \theta}{\partial z} \delta z'$$

Suppose a saturated parcel which has the environmental potential temperature at $z_0 - \delta z'$ is raised to the level z_0. When it arrives at z_0 the parcel will have potential temperature

$$\theta_1 = \left(\theta_0 - \frac{\partial \theta}{\partial z} \delta z'\right) + \delta \theta \qquad (12.19)$$

where $\delta \theta$ is the change in potential temperature due to ascent through vertical distance $\delta z'$. Assuming a pseudo-adiabatic ascent we see from (12.16) that

$$\frac{\delta \theta}{\theta} \simeq -\delta\left(\frac{L_c q_s}{c_p T}\right) \simeq -\frac{\partial}{\partial z}\left(\frac{L_c q_s}{c_p T}\right) \delta z'$$

so that the buoyancy of the parcel when it arrives at z_0 is proportional to

$$\frac{\theta_1 - \theta_0}{\theta_0} \simeq -\left[\frac{1}{\theta}\frac{\partial \theta}{\partial z} + \frac{\partial}{\partial z}\left(\frac{L_c q_s}{c_p T}\right)\right] \delta z' \qquad (12.20)$$

or with the aid of (12.18)

$$\frac{\theta_1 - \theta_0}{\theta_0} \simeq -\frac{\partial \ln \theta_e}{\partial z} \delta z'$$

Now, the saturated parcel will be positively buoyant at z_0 provided $\theta_1 > \theta_0$. Thus the conditional stability criterion for a saturated parcel is

$$\frac{\partial \theta_e}{\partial z} \begin{cases} < 0 & \text{conditionally unstable} \\ = 0 & \text{saturated neutral} \\ > 0 & \text{absolutely stable} \end{cases} \qquad (12.21)$$

In Fig. 12.1 the vertical profiles of θ and θ_e for a typical tropical sounding are shown. It is obvious from the figure that the mean tropical atmosphere is conditionally unstable in the lower troposphere. However, this observed profile does not imply that convective overturning will spontaneously occur in the tropics. The release of conditional instability requires not only $\partial \theta_e / \partial z < 0$, but also a saturated atmosphere, and the mean relative humidity in the tropics is well below 100 percent. Thus, low-level convergence with its resultant forced ascent is required to produce saturation. The amount of forced ascent necessary to produce a positively buoyant parcel can be estimated simply from Fig. 12.1. A parcel rising pseudo-adiabatically from a level $z_0 - \delta z'$ will conserve the value of θ_e characteristic of the environment at $z_0 - \delta z'$. Now, the buoyancy of a parcel depends only on the difference in density between the parcel and the environment. Thus, in order to compute the buoyancy of the parcel at z_0, it is not correct simply to compare θ_e of the environment at z_0 to $\theta_e(z_0 - \delta z')$ because if the environment is unsaturated the difference in θ_e for the parcel and the environment may be due primarily to

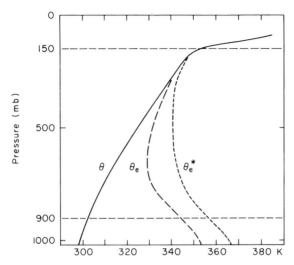

Fig. 12.1 Typical sounding in the tropical atmosphere showing the vertical profiles of potential temperature θ, equivalent potential temperature θ_e, and the equivalent potential temperature of a hypothetically saturated atmosphere with the same temperature at each level. (After Ooyama, 1969.)

the difference in mixing ratios, not to any temperature (density) difference. To estimate the buoyancy of the parcel $\theta_e(z_0 - \delta z')$ should instead be compared to $\theta_e^*(z_0)$ which is the equivalent potential temperature which the environment at z_0 would have if it were isothermally brought to saturation. The parcel will thus become buoyant when $\theta_e(z_0 - \delta z') > \theta_e^*(z_0)$, for then the parcel temperature will exceed the temperature of the environment. From Fig. 12.1 we see that θ_e for a parcel raised from 1000 mb will intersect the θ_e^* curve just above 900 mb, whereas a parcel raised from any level much above 900 mb will not interest θ_e^* no matter how far it is forced to ascend. It is for this reason that *low-level convergence* is required to initiate convective overturning in the tropical atmosphere. Only air near the surface has a sufficiently high value of θ_e to become buoyant when it is forcibly raised. Of course, convergence at higher levels may play an important role in maintaining the convection by adding substantial moisture to the system.

12.2.3 THE SLICE METHOD

In the parcel method the environment is assumed to remain undisturbed by the convective parcels. In reality the rising motion in the convection must be compensated by subsidence in the environment if an overall mass balance is to be maintained. For this reason, the parcel method tends to overestimate the degree of instability of the atmosphere. The simplest manner in which to take account of this adjustment in the environment is the so-called *slice method*. In this method it is assumed that on any horizontal plane the upward mass flux in the convection cells is just balanced by downward mass flux in the environment immediately surrounding the convection cells. This mass balance may be expressed as follows: We let ρ' and w' be the density and vertical velocity in the convection cells and ρ and w be the corresponding fields in the environment. If the fractional horizontal area occupied by the convection cells is designated by a, then for mass balance

$$\rho'w'a = \rho w(1 - a) \tag{12.22}$$

Now suppose that in a time increment δt the convective parcels rise an amount $\delta z'$ and the environment sinks by δz. We can write

$$w' \simeq \frac{\delta z'}{\delta t}, \qquad w \simeq \frac{\delta z}{\delta t}$$

Substituting into (12.22) and neglecting the small difference between ρ' and ρ we obtain

$$a\,\delta z' \simeq (1 - a)\,\delta z \tag{12.23}$$

We now consider the difference at a level z_0 between the temperatures of a saturated parcel which has risen from the level $z_0 - \delta z'$ and the environmental air which has sunk from the level $z_0 + \delta z$. Letting θ_0 be the initial undisturbed environmental potential temperature at z_0, we have for the potential temperature of the environment at level $z_0 + \delta z$

$$\theta = \theta_0 + \frac{\partial \theta}{\partial z}\,\delta z$$

The descending environmental air heats dry adiabatically. It thus preserves its original potential temperature and arrives at z_0 with potential temperature

$$\theta_2 = \theta_0 + \frac{\partial \theta}{\partial z}\,\delta z \tag{12.24}$$

The rising parcel, on the other hand, arrives at z_0 with potential temperature θ_1 given by (12.19). The buoyancy at z_0 is thus proportional to $\theta_1 - \theta_2$ which as shown in Fig. 12.2 is a smaller buoyancy than the $\theta_1 - \theta_0$ given by the parcel method. From (12.19) and (12.24) we find that

$$\frac{\theta_1 - \theta_2}{\theta} = -\frac{\partial \ln \theta_e}{\partial z}\,\delta z' - \frac{\partial \theta}{\partial z}\,\delta z \tag{12.25}$$

Therefore, with the aid of (12.23) we obtain the stability criterion

$$\frac{\partial \ln \theta_e}{\partial z} + \frac{\partial \ln \theta}{\partial z}\left(\frac{a}{1 - a}\right) \begin{cases} < 0 & \text{conditionally unstable} \\ = 0 & \text{neutral} \\ > 0 & \text{stable} \end{cases} \tag{12.26}$$

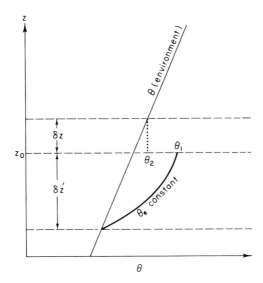

Fig. 12.2   Graphical computation of parcel buoyancy using the slice method.

Observations indicate that in general active convection cells occupy only a small percent of the total area even in synoptic disturbances. Thus, $a/(1 - a)$ $\ll 1$ and the slice stability does not in practice differ much from the ordinary moist parcel stability.

12.2.4   ENTRAINMENT

In the previous subsection it was assumed that the rising convective parcels do not mix with the environment. In reality, however, rising saturated parcels tend to be diluted by *entraining*, or mixing in, some of the relatively dry environmental air. If the air in the environment is unsaturated, some of the liquid water in the rising parcel must be evaporated to maintain saturation in the convection cell as air from the environment is entrained. The evaporative cooling caused by entrainment will reduce the buoyancy of the convective parcel. Thus, the equivalent potential temperature in an entraining convection cell will decrease with height rather than remaining constant. The magnitude of the entrainment effect can be determined by application of the thermodynamic energy equation to a mass M of saturated cloud air and a mass dM of entrained air. In entropy form the appropriate approximate energy equation for the cloud air alone is (12.16). Evaporation of liquid water to bring the entrained air to saturation will cause a diabatic heating of $-(q_s - \bar{q})L_c \, dM$, where $\bar{q}$ is the specific humidity of the environment. In addition an amount of heat, $c_p(T - \bar{T})dM$ is required to raise the temperature of the environmental air $\bar{T}$ to the cloud temperature. Thus the approximate thermodynamic energy equation for this case is

$$-Md\left(\frac{L_c q_s}{T}\right) - \frac{(q_s - \bar{q})L_c\,dM}{T} = Mc_p\,d\ln\theta + c_p\,\frac{(T - \bar{T})}{T}\,dM \quad (12.27)$$

where we have neglected the heat accession by the liquid water. From the definition of θ_e (12.18) we can write (12.27) as

$$-M\,d\ln\theta_e = dM\left[(q_s - \bar{q})\frac{L_c}{c_p T} + \frac{(T - \bar{T})}{T}\right]$$

or

$$\frac{\partial\ln\theta_e}{\partial z} = -\frac{d\ln M}{dz}\left[\frac{L_c}{c_p T}(q_s - \bar{q}) + \frac{T - \bar{T}}{T}\right] \quad (12.28)$$

For there to be positive buoyancy $T > \bar{T}$, and since $q_s \geq \bar{q}$ the terms in square brackets will be positive and if the entrainment is positive $(d\ln M/dz > 0)$ the equivalent potential temperature of the cloud air must decrease with height. It is observed that in typical cumulus clouds $T - \bar{T} \sim 1°C$. Thus, the first term in the brackets in (12.28) dominates. If the last term is neglected, (12.28) yields the maximum entrainment possible to just maintain neutral buoyancy for given profiles of θ_e, q_s, and $\bar{q}$.

Under certain restrictive assumptions, it is possible to use (12.28) together with the vertical momentum equation and continuity equation to predict the vertical temperature profile and maximum height of penetration for a cumulus cell given the temperature and specific humidity profiles in the environment. If the convection cell is assumed to be a steady-state entraining jet, then the momentum balance in a slice of depth δz and cross-sectional area S as shown in Fig. 12.3 may be computed as follows: We let

$$M = S\rho w \quad (12.29)$$

be the vertical mass flux at the level $z = z_0$. Then the vertical mass flux at $z = z_0 + \delta z$ may be written

$$M + \delta M = M + \frac{M}{dz}\,\delta z$$

The vertical *momentum* flux is thus wM at z_0 and

$$\left(w + \frac{dw}{dz}\,\delta z\right)\left(M + \frac{dM}{dz}\,\delta z\right)$$

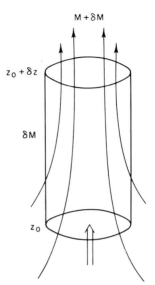

Fig. 12.3 An entraining jet model of cumulus convection.

at $z_0 + \delta z$. Thus, the net divergence of momentum from the volume element $S \, \delta z$ is

$$\left(w + \frac{dw}{dz} \delta z \right) \left(M + \frac{dM}{dz} \delta z \right) - wM \simeq \frac{d}{dz} (wM) \, \delta z$$

For steady-state motion, the momentum divergence must be balanced by production of vertical momentum by the buoyancy force. It was shown in Chapter 9 [see Eq. (9.18)] that the buoyancy force per unit mass is approximately

$$g \frac{(T - \bar{T})}{T}$$

Thus, the momentum balance for the mass $\rho S \, \delta z$ becomes

$$\frac{d}{dz} (wM) \, \delta z = g \frac{(T - \bar{T})}{T} S \rho \, \delta z$$

or after eliminating S with (12.29)

$$\frac{d}{dz} (wM) = \frac{M}{w} g \frac{(T - \bar{T})}{T} \tag{12.30}$$

Equation (12.30) can also be written in the form

$$\frac{1}{2} \frac{d}{dz} (w^2) = g \left(\frac{T - \bar{T}}{T} \right) - \frac{d \ln M}{dz} w^2 \tag{12.31}$$

which indicates that the rate of kinetic energy increase with height is given by the kinetic energy production by buoyancy minus the kinetic energy loss due to entrainment.

Provided that the cross section S of the convective cell is known as a function of height the system (12.28)–(12.30) may be regarded as a simultaneous set of equations for the determination of the vertical profiles of M, w, and T given $\bar{q}$ and $\bar{T}$. This model can to some extent duplicate observed conditions in cumulus clouds; however, it by no means constitutes a complete theory of cumulus convection because the cross section S must be empirically specified. In addition the entraining jet model assumes a steady state which is certainly not valid for real cumulus convection.

12.2.5 HEATING BY CONDENSATION

The manner in which the atmosphere is heated by condensation of water vapor depends crucially on the nature of the condensation process. In particular, it is necessary to differentiate between latent heat release through large-scale vertical motion (that is, synoptic scale forced uplift) and the latent heat release due to small-scale deep cumulus convection. The former process, which is generally associated with midlatitude synoptic systems, can be easily incorporated into the thermodynamic energy equation in terms of the synoptic scale field variables. However, since the large scale heating field resulting from the cooperative action of many cumulonimbus cells involves a complex interaction between the cumulus scale and synoptic scale motions, representation of this type of latent heating in terms of the synoptic scale field variables is much more difficult.

Before considering the problem of condensation heating by cumulus convection, it is worth indicating briefly how the condensation heating by large-scale forced uplift can be included in a prediction model. The approximate thermodynamic energy equation for a pseudo-adiabatic process (12.15) states that

$$\frac{d \ln \theta}{dt} \simeq - \frac{L_c}{c_p T} \frac{dq_s}{dt} \tag{12.32}$$

Now the change in q_s following the motion is primarily due to ascent, so that

$$\frac{dq_s}{dt} \simeq w \frac{\partial q_s}{\partial z} \qquad \text{for} \quad w > 0$$

$$\frac{dq_s}{dt} \simeq 0 \qquad \qquad \text{for} \quad w < 0 \tag{12.33}$$

and (12.32) can be written in the form

$$\left(\frac{\partial}{\partial t} + \mathbf{V} \cdot \nabla\right) \ln \theta + w \left[\frac{\partial \ln \theta}{\partial z} + \frac{L_c}{c_p T} \frac{\partial q_s}{\partial z}\right] \simeq 0 \qquad (12.34)$$

for regions where $w > 0$. But from (12.18)

$$\frac{\partial \ln \theta_e}{\partial z} \simeq \frac{\partial \ln \theta}{\partial z} + \frac{L_c}{c_p T} \frac{\partial q_s}{\partial z}$$

so that (12.34) can be written in a form valid for both positive and negative vertical motion as

$$\left(\frac{\partial}{\partial t} + \mathbf{V} \cdot \nabla\right) \theta + w\Gamma_e \simeq 0 \qquad (12.35)$$

where Γ_e is an *equivalent static stability* defined by

$$\Gamma_e = \frac{\theta}{\theta_e} \frac{\partial \theta_e}{\partial z} \qquad \text{for} \quad q > q_s \quad \text{and} \quad w > 0$$

$$\Gamma_e = \frac{\partial \theta}{\partial z} \qquad \text{for} \quad q < q_s \quad \text{or} \quad w < 0$$

Thus, in the case of condensation due to large scale forced ascent ($\Gamma_e > 0$) the thermodynamic energy equation has essentially the same form as for adiabatic motions except that the static stability is replaced by the equivalent static stability. As a consequence, the local temperature changes induced by the forced ascent will be smaller than for the case of forced dry ascent with the same lapse rate.

If, on the other hand, $\Gamma_e < 0$, the atmosphere is conditionally unstable and the condensation will occur primarily through cumulus convection. In that case (12.33) is still valid but the vertical velocity must be that of the individual cumulus updrafts, not the synoptic scale w. Thus, a simple formulation of the thermodynamic energy equation in terms of only the synoptic scale variables is not possible. However, we can still simplify the thermodynamic energy equation to some extent. We recall from Section 12.1 [see Eq. (12.11)] that due to the smallness of temperature fluctuations in the tropics the adiabatic cooling and diabatic heating terms must approximately balance. Thus, (12.32) becomes approximately

$$w \frac{\partial \ln \theta}{\partial z} = -\frac{L_c}{c_p T} \frac{dq_s}{dt} \qquad (12.36)$$

The synoptic scale vertical velocity w which appears in (12.36) is the average of very large vertical motions in the active convection cells and small vertical motions in the environment. Thus, if we let w' be the vertical velocity in the convective cells and $\bar{w}$ the vertical velocity in the environment, we have

$$w = aw' + (1 - a)\bar{w} \tag{12.37}$$

where a is the fractional area occupied by the convection. With the aid of (12.33) we can then write (12.36) in the form

$$w \frac{\partial \ln \theta}{\partial z} = -\frac{L}{c_p T} aw' \frac{dq_s}{\partial z} \tag{12.38}$$

The problem is then to express the condensation heating term on the right in (12.38) in terms of the synoptic scale field variables. A number of empirical models for doing so have been suggested; however, none of these is completely satisfactory from a theoretical viewpoint.

12.3   The Observed Structure of Large-Scale Motions in the Equatorial Zone

In the first section of this chapter we have presented scaling arguments which indicate that the large-scale motions in the equatorial zone are driven primarily by latent heat release. Observations indicate that this latent heating occurs primarily through cumulonimbus convection rather than large-scale forced ascent. For this type of driving force to be effective, an interaction between the cumulus scale and large-scale motions is necessary in which the large-scale convergence provides moisture for the convection and the cumulus cells act cooperatively to provide a large-scale heat source. Because of the special nature of this driving force, as well as the smallness of the Coriolis parameter, large-scale equatorial motion systems have certain distinctive characteristic structural features which are quite different from those of midlatitude systems.

12.3.1   THE INTERTROPICAL CONVERGENCE ZONE

In the traditional view the tropical general circulation was thought to consist of a direct Hadley circulation in which the air in the lower troposphere in both hemispheres moved equatorward toward the *intertropical convergence zone* (ITCZ) where by continuity considerations it was forced to rise and move poleward, thus transporting heat away from the equator in the upper troposphere. However, this simple model of large-scale overturning is not consistent with the observed vertical profile of θ_e. As indicated in Fig. 12.1, the equivalent static stability in the tropics is positive above about 600 mb. Thus, a large scale upward mass flow would be up the gradient of θ_e in the upper troposphere and would actually cool the upper troposphere in the region of the ITCZ [see Eq. (12.35)]. Such a circulation could not generate potential energy and would not, therefore, satisfy the heat balance in the equatorial zone.

In fact it appears that the only way in which heat can effectively be brought from the surface to the upper troposphere in the ITCZ is through pseudo-adiabatic ascent in the cores of large cumulonimbus clouds (often referred to as "hot towers"). For such motions, the air parcels approximately conserve θ_e and can therefore arrive in the upper troposphere with a moderate tempera-ture excess. Thus, the heat balance of the equatorial zone can be accounted for, at least qualitatively, provided that the vertical motion in the ITCZ is confined primarily to the updrafts of individual convective cells. Riehl and Malkus (1968) have estimated that only 1500–5000 individual "hot towers" would need to exist simultaneously around the globe to account for the required vertical heat transport in the ITCZ.

This view of the ITCZ as a narrow zonal band of vigorous cumulus convection has been confirmed beyond doubt by recent observations, par-ticularly satellite cloud photos. Observations indicate that within the ITCZ precipitation greatly exceeds the moisture supplied by evaporation from the ocean surface below. Thus, much of the vapor necessary to maintain the convection in the ITCZ must be supplied by the converging tradewind flow in the lower troposphere. In this manner the large-scale flow provides the latent heat supply for the convection and the convective heating in turn produces the large-scale pressure field which maintains the low-level inflow.

The above description of the ITCZ is actually oversimplified. In reality, the ITCZ over the oceans rarely appears as a long unbroken band of heavy convective cloudiness, and it almost never is found right at the equator. Rather, it is usually made up of a number of distinct *cloud clusters*, with scales of the order of a few hundred kilometers, which are separated by regions of relatively clear skies. The strength of the ITCZ is also quite variable in both space and time. It is particularly persistent and well defined over the Pacific and Atlantic between about 5 and 10°N latitude, and occasionally appears in the Pacific between 5 and 10°S. That the ITCZ is centered away from the equator is clearly demonstrated by Fig. 12.4 which shows a 15-day average of the satellite cloud brightness photographs for the Pacific. The dry zone along the equator is a particularly striking feature.

12.3.2 EQUATORIAL WAVE DISTURBANCES

The cloud clusters which are observed along the ITCZ are actually in general just precipitation zones associated with weak wave disturbances which propagate westward along the ITCZ. That such westward propagating disturbances exist and are responsible for a large part of the cloudiness in the

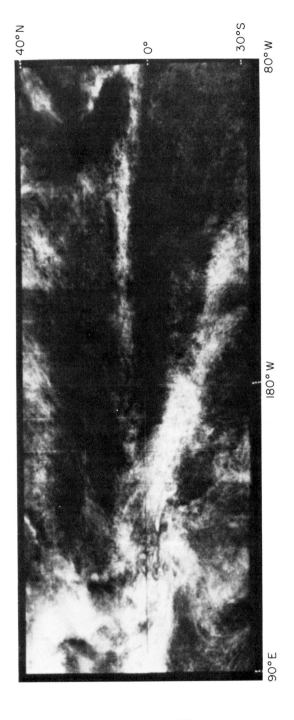

Fig. 12.4 Fifteen-day satellite cloud brightness averages for the equatorial Pacific, 16–31 January 1967 (Mercator projection). (After Kornfield *et al.*, 1967.)

ITCZ can be easily seen by viewing time–longitude sections constructed from daily satellite pictures cut into thin zonal strips. An example is shown in Fig. 12.5. The well-defined bands of cloudiness which slope from right to left down the page define the locations of the cloud clusters as a function of time. Clearly much of the cloudiness in the 5–10°N latitude zone of the Pacific is associated with westward moving disturbances. The slope of the cloud lines in Fig. 12.5 implies a westward propagation speed of about 8–10 m sec^{-1}. The longitudinal separation of the cloud bands is about 3000–4000 km, corresponding to a period range of about 4–5 days for this type of disturbance.

Evidence from diagnostic studies indicates that these westward propagating disturbances are driven by the release of latent heat in the convective precipitation areas accompanying the waves. The vertical structure of this type of disturbance is shown in schematic form in Fig. 12.6. Since the approximate thermodynamic energy equation (12.38) requires that the vertical motion be proportional to the diabatic heating, the maximum large-scale vertical velocities occur in the convective zone. By continuity there must thus be convergence and cyclonic vorticity generation at low levels in the convection zone and divergence with anticyclonic vorticity generation in the upper troposphere. Hence, the convection tends to generate a low-level trough and upper-level ridge.

The strongest convective activity in these waves occurs where the midtropospheric temperatures are warmer than average (although generally by less than 1°C). The correlation between temperature and the diabatic heating rate is therefore positive and from the approximate thermodynamic energy equation (12.12) we find that

$$\overline{w^*T\Gamma} \simeq \frac{\overline{TH}}{c_p} \tag{12.39}$$

where the overbar indicates an average over one wavelength. Thus, the potential energy generated by the diabatic heating is immediately *converted* to kinetic energy through the $\overline{w^*T}$ correlation. There is, in this approximation, no storage in the form of available potential energy. The energetics of these disturbances, therefore, differs remarkably from that of midlatitude baroclinic systems in which the available potential energy greatly exceeds the kinetic energy.

A typical vertical profile of divergence in the precipitation zone of this type of disturbance is shown in Fig. 12.7. Convergence is not limited to low-level frictional inflow in the planetary boundary layer, but extends up to nearly 400 mb which is the height where the hot towers achieve their maximum buoyancy. The deep convergence implies that there must be substantial entrainment of midtropospheric air into the convective cells. Because the

Fig. 12.5 Time longitude sections of satellite photographs for the period 1 July–14 August 1967 in the 5–10°N latitude band of the Pacific. The westward progression of the cloud clusters is indicated by the bands of cloudiness sloping down the page from right to left. (After Chang, 1970.)

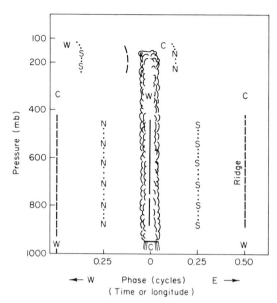

Fig. 12.6 Schematic model for equatorial wave disturbances showing trough axis (solid lines), ridge axis (dashed lines), axes of northerly and southerly wind components (dotted lines). Regions of warm and cold air designated by W and C, respectively. (After Wallace, 1971.)

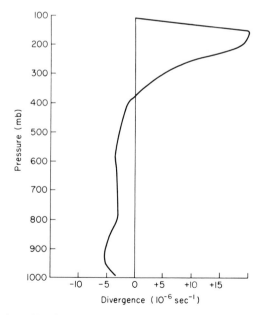

Fig. 12.7 Vertical profile of 4°-square area average divergence based on composites of many equatorial disturbances. (Adapted from Williams, 1969.)

midtropospheric air is relatively dry, this entrainment will require consider-able evaporation of liquid water to bring the mixture of cloud and environ-ment air to saturation. It will thus reduce the buoyancy of the cloud air, and may in fact produce negatively buoyant convective downdrafts if there is sufficient evaporative cooling. However, in the large cumulonimbus clouds present in equatorial waves the central core updrafts are protected from entrainment by the surrounding cloud air so that they can penetrate nearly to the tropopause without dilution by environmental air. It is these undiluted cores which constitute the "hot towers" referred to in the previous section. Since the hot towers are responsible for most of the vertical heat transport in the ITCZ, and the wave disturbances contain most of the active convective precipitation areas along the ITCZ, it is obvious that the equatorial waves play an essential role in the general circulation of the atmosphere.

12.3.3   WAVES IN THE EQUATORIAL STRATOSPHERE

In Section 11.7.3 we discussed the role of large-scale vertically prop-agating gravity waves in providing a zonal momentum source for the quasi-biennial oscillation. We also discussed the nature of pure internal gravity waves in Section 9.4.3. However, large-scale internal gravity waves differ somewhat from the simple pure gravity modes because they are influenced by the earth's rotation as well as the gravitational stability. Such waves are often referred to as *inertia–gravity* oscillations.

It can be shown that planetary scale waves generally will be "trapped" (that is, cannot propagate energy vertically) unless the frequency of the wave is greater than the Coriolis frequency. Thus at middle latitudes waves with periods in the range of several days are generally not able to propagate significantly into the stratosphere. However, as the equator is approached, the decreasing Coriolis frequency allows these longer period waves to become untrapped and propagate vertically.

The theory predicts that these long-period vertically propagating waves generally must have very short vertical wavelengths. Therefore, even if they existed such waves could not be resolved by conventional radiosonde data. However, there are two important types of vertically propagating waves whose wavelengths are long enough to permit easy detection. These are the eastward traveling atmospheric *Kelvin wave*, and the westward traveling mixed *Rossby–gravity wave*. The horizontal distributions of pressure and velocity character-istic of these waves is shown in Fig. 12.8. The Kelvin wave has a distribution of pressure and zonal velocity which is symmetric about the equator, and has essentially *no meridional velocity* component. The mixed Rossby–gravity wave, on the other hand, has a distribution of pressure and zonal velocity

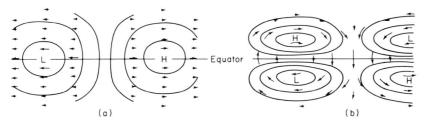

Fig. 12.8 Velocity and pressure distributions in the horizontal plane for (a) Kelvin waves, and (b) mixed Rossby–gravity waves. (After Matsuno, 1966.)

which is antisymmetric about the equator, and has a distribution of meridional velocity which is symmetric.

The dynamics of these waves can easily be deduced theoretically by applying the linear perturbation technique introduced in Chapter 9. It is convenient to use the governing equations in log-pressure coordinates (12.1)–(12.4) referred to an *equatorial β plane* in which the Coriolis parameter is approximated as

$$f = \frac{2\Omega y}{a} \equiv \beta y$$

Thus, β is the rate at which f changes with latitude at the equator. We assume as a basic state an atmosphere at rest with no diabatic heating. If the perturbations are assumed to be zonally propagating waves, we can write

$$
\begin{aligned}
u &= u'(y, z^*)e^{i(kx + vt)} \\
v &= v'(y, z^*)e^{i(kx + vt)} \\
w^* &= w^{*\prime}(y, z^*)e^{i(kx + vt)} \\
\Phi &= \Phi'(y, z^*)e^{i(kx + vt)}
\end{aligned}
\tag{12.40}
$$

Using (12.2) to eliminate temperature in (12.4) we obtain the following set of linearized perturbation equations:

$$ivu' - \beta y v' = -ik\Phi' \tag{12.41}$$

$$ivv' + \beta y u' = -\frac{\partial \Phi'}{\partial y} \tag{12.42}$$

$$iku' + \frac{\partial v'}{\partial y} + \left(\frac{\partial}{\partial z^*} - \frac{1}{H}\right)w^{*\prime} = 0 \tag{12.43}$$

$$iv\frac{\partial \Phi'}{\partial z^*} + \frac{R\Gamma}{H}w^{*\prime} = 0 \tag{12.44}$$

For the Kelvin-wave case, this set can be considerably simplified. Setting $v' \equiv 0$ and eliminating $w^{*'}$ between (12.43) and (12.44) we obtain

$$ivu' = -ik\Phi' \tag{12.45}$$

$$\beta yu' = -\frac{\partial \Phi'}{\partial y} \tag{12.46}$$

$$\left(\frac{\partial}{\partial z^*} - \frac{1}{H}\right)\frac{\partial \Phi'}{\partial z^*} - \frac{k}{v}Su' = 0 \tag{12.47}$$

where $S = R\Gamma/H$. Using (12.45) to eliminate Φ' in (12.46) and (12.47), we obtain two independent equations which the field of u' must satisfy:

$$\beta yu' = \frac{v}{k}\frac{\partial u'}{\partial y} \tag{12.48}$$

and

$$\left(\frac{\partial}{\partial z^*} - \frac{1}{H}\right)\frac{\partial u'}{\partial z^*} + \frac{k^2}{v^2}Su' = 0 \tag{12.49}$$

Equation (12.48) determines the meridional distribution of u' and (12.49) determines the vertical distribution.

It may be easily verified that (12.48) has the solution

$$u' = u_0(z^*)\exp\{\beta y^2 k/2v\} \tag{12.50}$$

If we assume that $k > 0$, then $v < 0$ corresponds to an eastward-moving wave. In that case (12.50) indicates that the field of u' will have a Gaussian distribution about the equator with an e-folding width given by

$$Y_{\mathrm{L}} = \left|\frac{2v}{\beta k}\right|^{1/2} \tag{12.51}$$

For a westward-moving wave ($v > 0$) the solution (12.50) increases in amplitude exponentially away from the equator. This solution cannot satisfy reasonable boundary conditions at the poles and must, therefore, be rejected. Therefore, there exists only an *eastward* propagating atmospheric Kelvin wave.

The equation for the vertical dependence of u' (12.49) is simply the vertical structure equation for an internal gravity wave discussed previously in Section 9.4.3. Solutions can be written in the form

$$u_0(z^*) = e^{z^*/2H}[C_1 e^{i\lambda z^*} + C_2 e^{-i\lambda z^*}] \tag{12.52}$$

with

$$\lambda^2 \equiv \frac{Sk^2}{v^2} - \frac{1}{4H^2}$$

Here the constants C_1 and C_2 are to be determined by appropriate boundary conditions. For $\lambda^2 > 0$ the solution (12.52) is in the form of a vertically propagating wave. For waves in the equatorial stratosphere which are forced by disturbances in the troposphere, the energy propagation (that is, group velocity) must have an upward component. Therefore, according to the arguments of Section 9.4.4 the phase velocity must have a downward component. Hence, the constant $C_1 = 0$ in (12.52) and, the Kelvin wave has a structure in the x, z plane which is identical to that of the eastward-propagating pure internal gravity wave shown in Fig. 9.3. In summary, the atmospheric Kelvin wave is an eastward-moving wave in which the zonal velocity and meridional pressure fields are in exact geostrophic balance so that on the equatorial β-plane the meridional velocity is identically zero and the wave can propagate as an ordinary two-dimensional internal gravity wave in the x, y plane.

A similar analysis is possible for the mixed Rossby–gravity mode. For this case, we must use the full perturbation equations (12.41)–(12.44). After considerable algebraic manipulation, it can be shown that the solution corresponding to the pattern shown in Fig. 12.8 is

$$\begin{Bmatrix} u' \\ v' \\ \Phi \end{Bmatrix} = \Psi(z^*) \begin{Bmatrix} -i\beta y(1 - kv/\beta)/v \\ 1 \\ -ivy \end{Bmatrix} \exp\left[\frac{-(1 - kv/\beta)\beta^2 y^2}{2v^2}\right] \qquad (12.53)$$

where the vertical structure $\Psi(z^*)$ of the three variables is given by

$$\Psi(z^*) = e^{z^*/2H}[C_1 e^{i\lambda_0 z^*} + C_2 e^{-i\lambda_0 z^*}] \qquad (12.54)$$

with

$$\lambda_0{}^2 \equiv S \frac{k^2}{v^2}\left(1 - \frac{\beta}{vk}\right)^2 - \frac{1}{4H^2}$$

and the constants C_1 and C_2 are again to be determined by the boundary conditions.

It is obvious from (12.53) that for the mixed Rossby–gravity mode v' has a Gaussian distribution about the equator. The e-folding width of the oscillation is in this case

$$Y_L = \left|\frac{2v}{\beta(\beta/v - k)}\right|^{1/2} \qquad (12.55)$$

This solution is valid for westward propagating waves ($v > 0$) provided that

$$1 - \frac{kv}{\beta} > 0$$

or noting that $s = ka$ where s is the number of wavelengths around a latitude circle this condition may be written as

$$v < \frac{2\Omega}{s} \tag{12.56}$$

For frequencies which do not satisfy (12.56), the wave amplitude will not decay away from the equator and it is not possible to satisfy boundary conditions at the pole.

Both Kelvin wave and mixed Rossby–gravity wave modes have been identified in observational data from the equatorial stratosphere. The observed Kelvin waves have periods in the range of 12–20 days and appear to be primarily of zonal wavenumber $s = 1$ (that is, one wavelength spans $360°$ longitude). The corresponding observed phase speeds of these waves are in the range of $v/k \sim 30$ m sec^{-1}. Knowing the phase speed we can use (12.51) to compute a predicted lateral width for the observed waves of $Y_L \simeq 1800$ km. This agrees well with observational indications that the Kelvin waves have significant amplitude only within about $20°$ of the equator. Knowledge of the observed phase speed also allows us to compute the theoretical vertical wavelength of the Kelvin wave. Assuming that the stratosphere is isothermal, we obtain from (12.52)

$$\frac{2\pi}{\lambda} \simeq S^{-1/2} \frac{v}{k} \simeq 12 \quad \text{km}$$

which agrees well with the vertical wavelength deduced from observations.

An example of zonal wind oscillations caused by the passage of Kelvin waves at a station near the equator is shown in the form of a time–height section in Fig. 12.9. During the observational period shown in the figure the westerly phase of the quasi-biennial oscillation is descending so that at each level there is a general increase in the mean zonal wind with increasing time. However, superposed on this secular trend is a large fluctuating component with a period between speed maxima of about 12 days and a vertical wavelength (computed from the tilt of the oscillations with height) of about 10–12 km. Observations of the temperature field for the same period reveal that the temperature oscillation leads the zonal wind oscillation by one-quarter cycle (that is, maximum temperature occurs one-quarter period prior to maximum westerlies) which is just the phase relationship required for Kelvin waves (see Fig. 9.3). Furthermore, additional observations from other stations indicate that these oscillations do propagate eastward at about 30 m sec^{-1}. Therefore there can be little doubt that the observed oscillations are Kelvin waves.

The existence of the mixed Rossby–gravity mode has also been confirmed

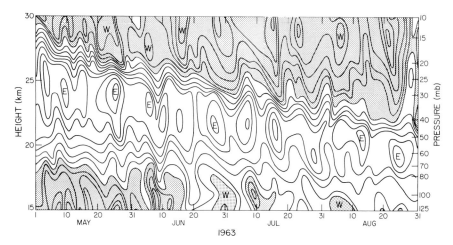

Fig. 12.9 Time–height section of zonal wind at Kwajalein (9°N latitude). Isotachs at 5 m sec^{-1} intervals. Westerlies are shaded. (After Wallace and Kousky, 1968.)

in observational data from the equatorial Pacific. This mode is most easily identified in the meridional wind component, since from (12.53) v' is a maximum at the equator for the mixed Rossby–gravity mode. The observed waves of this mode have periods in the range of 4–5 days and propagate westward at about 20 m sec^{-1}. The horizontal wavelength appears to be about 10,000 km, corresponding to zonal wave number $s = 4$. The observed vertical wavelength is about 6 km, which agrees with the theoretically derived wavelength (12.54) within the uncertainties of the observations. These waves also appear to have significant amplitude only within about 20° of the equator which is consistent with the theoretical e-folding width (12.55).

At present it appears that both the Kelvin waves and the mixed Rossby–gravity waves are excited by oscillations in the large-scale convective heating pattern in the equatorial troposphere. Although these waves do not contain much energy compared to typical tropospheric disturbances, they are the predominate disturbances of the equatorial stratosphere, and through their vertical energy and momentum transport play a crucial role in the general circulation of the stratosphere.

12.4 The Origin of Equatorial Disturbances

Some tropical disturbances probably originate as midlatitude baroclinic waves which move equatorward and gradually assume tropical characteristics. However, there can be little doubt that most synoptic scale disturbances in the equatorial zone originate *in situ*. Baroclinic instability cannot account for the development of these disturbances because, as we have shown from scaling arguments, the available potential energy is very small in the equatorial zone. Two processes which have been suggested as possible mechanisms for initiating equatorial disturbances are *barotropic instability* of the mean zonal flow due to lateral shear, and *conditional instability of the second kind* (CISK) due to organized convection driven by frictional moisture convergence in the boundary layer. Neither of these mechanisms can be readily tested with a simple analytic normal modes stability analysis such as the one which we carried out for baroclinic instability. Therefore, we are limited here to a fairly qualitative discussion.

12.4.1 BAROTROPIC INSTABILITY

According to the scale analysis of Section 12.1, in the absence of condensation vertical motions must be small in the tropics. Thus, to a first approximation the flow is governed by the barotropic vorticity equation

$$\left(\frac{\partial}{\partial t} + \mathbf{V} \cdot \mathbf{V}\right)(\zeta + f) = 0 \tag{12.57}$$

We now assume that the flow consists of a small barotropic perturbation superposed on a zonal current which depends only on latitude. Thus, we let

$$u = \bar{u}(y) + u', \qquad v = v'$$

Since the flow is quasi-non-divergent, we can express the perturbed motion in terms of a perturbation streamfunction ψ' by letting

$$u' = -\frac{\partial \psi'}{\partial y}, \qquad v' = \frac{\partial \psi'}{\partial x}$$

we can then write the linearized perturbation form of (12.57) as

$$\left(\frac{\partial}{\partial t} + \bar{u}\frac{\partial}{\partial x}\right)\nabla^2 \psi' + \left(\beta - \frac{d^2\bar{u}}{dy^2}\right)\frac{\partial \psi'}{\partial x} = 0 \tag{12.58}$$

The quantity

$$\beta - \frac{d^2\bar{u}}{dy^2} = \frac{d}{dy}(f + \bar{\zeta})$$

is merely the latitudinal gradient of the basic state absolute vorticity. We now assume following the technique introduced in Chapter 9 that solutions of (12.58) can be represented in terms of zonally propagating harmonic waves in the form

$$\psi'(x, y, t) = \psi(y)e^{ik(x - ct)} \tag{12.59}$$

where $\psi = \psi_r = i\psi_i$ is a complex function of y alone. Substituting from (12.59) into (12.58) we obtain

$$(\bar{u} - c)\left(\frac{d^2\psi}{dy^2} - k^2\psi\right) + \left(\beta - \frac{d^2\bar{u}}{dy^2}\right)\psi = 0 \qquad (12.60)$$

which is an ordinary second-order differential equation in $\psi(y)$. As boundary conditions for (12.60), it is usually assumed that the perturbed flow is confined to a zonal channel with walls at $y = \pm L$ so that

$$\psi(y) = 0 \qquad \text{at} \quad y = \pm L \qquad (12.61)$$

The presence of the physically unrealistic walls at $y = \pm L$ should not significantly affect the results for perturbations whose amplitudes are small near the walls. For a given distribution of $\bar{u}(y)$ it turns out that (12.60) will have solutions which satisfy (12.61) only for certain values of the phase speed c. In cases where c is complex with a positive imaginary part we see from (12.59) that the perturbation amplitude will grow exponentially in time.

In practice it is not a simple matter to determine the solutions of (12.60) for a particular profile $u(y)$ because the coefficients are not constant. However, it is possible to obtain *necessary* conditions for the existence of instability by application of simple integral considerations. Dividing through in (12.60) by $(\bar{u} - c)$ we have

$$\frac{d^2\psi}{dy^2} - \left(k^2 - \frac{\beta - d^2\bar{u}/dy^2}{\bar{u} - c}\right)\psi = 0 \qquad (12.62)$$

If the phase speed is complex, $c = c_r + ic_i$, then $(\bar{u} - c)^{-1}$ is also complex and has real and imaginary parts

$$\delta_r = \frac{u - c_r}{(u - c_r)^2 + c_i^2}$$

$$\delta_i = \frac{c_i}{(u - c_r)^2 + c_i^2}$$

Equation (12.62) can then be separated into real and imaginary parts as follows:

$$\frac{d^2\psi_r}{dy^2} - \left[k^2 - \left(\beta - \frac{d^2\bar{u}}{dy^2}\right)\delta_r\right]\psi_r - \left(\beta - \frac{d^2\bar{u}}{dy^2}\right)\delta_i\psi_i = 0 \qquad (12.63)$$

$$\frac{d^2\psi_i}{dy^2} - \left[k^2 - \left(\beta - \frac{d^2\bar{u}}{dy^2}\right)\delta_r\right]\psi_i + \left(\beta - \frac{d^2\bar{u}}{dy^2}\right)\delta_i\psi_r = 0 \qquad (12.64)$$

Multiplying (12.63) by ψ_i, (12.63) by ψ_r, and subtracting the latter from the former, we obtain

$$\psi_i \frac{d^2\psi_r}{dy^2} - \psi_r \frac{d^2\psi_i}{dy^2} - \left(\beta - \frac{d^2\bar{u}}{dy^2}\right)\delta_i(\psi_i^2 + \psi_r^2) = 0$$

which can also be written as

$$\frac{d}{dy}\left(\psi_i \frac{d\psi_r}{dy} - \psi_r \frac{d\psi_i}{dy}\right) = \left(\beta - \frac{d^2\bar{u}}{dy^2}\right)\delta_i(\psi_r^2 + \psi_i^2) \qquad (12.65)$$

Integrating (12.65) with respect to y and applying the boundary conditions

$$\psi_r = \psi_i = 0 \qquad \text{at} \quad y = \pm L$$

we find that the terms on the left integrate to zero so that we are left with the integral condition

$$\int_{-L}^{+L} \left(\beta - \frac{d^2\bar{u}}{dy^2}\right)\delta_i |\psi|^2 \, dy = 0 \qquad (12.66)$$

where

$$|\psi|^2 = \psi_r^2 + \psi_i^2$$

Now, for an unstable perturbation to exist requires that $\delta_i > 0$. Since $|\psi|^2 \geq 0$ everywhere in the domain, the integral condition in (12.66) can be satisfied for an unstable wave only if $\beta - d^2\bar{u}/dy^2$ *changes sign* somewhere in the region $-L < y < L$. Thus, a *necessary* condition for barotropic instability is that the gradient of absolute vorticity of the mean current must vanish somewhere in the region, that is,

$$\beta - \frac{d^2\bar{u}}{dy^2} = 0 \qquad \text{somewhere} \qquad (12.67)$$

In Fig. 12.10 some observed profiles of monthly mean zonal wind and absolute vorticity as functions of latitude are shown for various pressure levels in the tropical Western Pacific. It is apparent that the *necessary* condition for barotropic instability is satisfied at some levels in the vicinity of the ITCZ in both months shown. Numerical solutions of the perturbation equation (12.62) for various profiles of $\bar{u}(y)$ indicate that the necessary condition which we have derived above is in general a sufficient condition for the existence of barotropic instability. Typically, the wavelength corresponding to maximum instability is on the order of 2000 km, which is comparable to the observed wavelengths for synoptic disturbances along the ITCZ.

Thus, barotropic instability appears to be a plausible mechanism for the development of weak disturbances along the ITCZ. However, barotropically

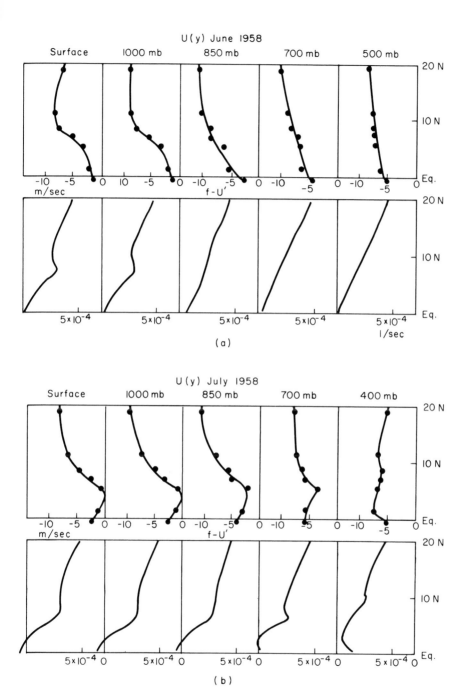

Fig. 12.10 Mean zonal wind and absolute vorticity profiles observed at various levels over the Marshall Islands in June and July 1958. (After Nitta and Yanai, 1969.)

unstable disturbances will continue to grow only if the shear of the mean zonal flow remains unstable so that the waves can extract energy from the mean flow. Since tropical disturbances are observed to exist in the absence of strong lateral shear it is unlikely that barotropic instability is the primary energy source for the maintenance of the waves once they have developed beyond their initial small-amplitude stage. Finally, it should be mentioned that barotropic instability is not a uniquely tropical phenomenon. The Coriolis parameter only appears in the barotropic vorticity equation in differentiated form as β. Thus, the equator has no special significance for barotropic flow. Barotropic instability is also possible in the vicinity of the midlatitude jet stream. However, at middle latitudes baroclinic instability is generally the more important mechanism.

12.4.2 CONDITIONAL INSTABILITY OF THE SECOND KIND

As we have previously pointed out, the release of latent heat in cumulus convection is almost certainly the primary energy source for the maintenance of finite-amplitude equatorial wave disturbances. It might be thought, therefore, that such waves are merely a direct result of the release of conditional instability in a saturated atmosphere with a super moist-adiabatic lapse rate ($\partial \theta_e/\partial z < 0$). However, numerous theoretical studies have shown that conditional instability produces maximum growth rates for motions on the scale of individual cumulus clouds. Therefore, ordinary conditional instability cannot explain the synoptic scale organization of the motion. Observations indicate, moreover, that the mean tropical atmosphere is not saturated even in the planetary boundary layer. Thus, a parcel must undergo a considerable amount of forced ascent before it becomes positively buoyant. Such forced ascent will occur only in regions of low level convergence. The cumulus convection and the large-scale motion must then be viewed as cooperatively interacting. The cumulus supplies the heat necessary to drive the large scale disturbance, and the large scale disturbance produces the moisture convergence necessary to drive the cumulus convection. This interaction process is indicated schematically in Fig. 12.11.

When the cooperative interaction between the cumulus convection and a large-scale perturbation leads to unstable growth of the large scale system the process is referred to as *conditional instability of the second kind*,[1] or CISK. The primary difficulty in testing the CISK mechanism theoretically is that the heating field due to the cumulus clouds must be expressed in terms of the large-scale field variables. Otherwise, it would be necessary to model the

---

[1] This term was coined by Charney and Eliassen (1964). The acronym CISK was first used by Rosenthal (1968).

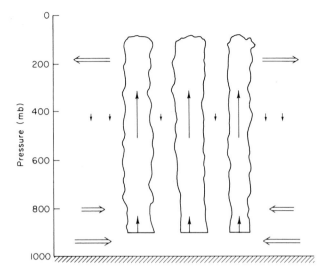

Fig. 12.11 The relationship between forced convection and low-level synoptic scale convergence.

detailed structure of all the cumulus elements in the disturbance. Several methods of doing this have been suggested. The results as applied to the stability of equatorial waves have so far been inconclusive. The CISK mechanism has, however, been applied with some success in theoretical studies of the growth of tropical cyclones. This subject will be touched on briefly in the next section.

12.5 Tropical Cyclones

Tropical cyclones are intense vortical storms which develop over the tropical oceans in regions of very warm surface water. These storms, which are called hurricanes in the Atlantic and typhoons in the Pacific, owe their existence to a very effective interactive coupling between the cumulus scale motion and the large-scale wind field.

In tropical cyclones the horizontal scale of the region where convection is strong is typically about 100 km in radius. Maximum tangential wind speeds in these storms range typically from 50–100 m sec^{-1}. For such high velocities and relatively small scales, the centrifugal force term cannot be neglected compared to the Coriolis force. Thus, to a first approximation the radial force balance in a steady-state hurricane satisfies the gradient wind relationship, not geostrophic balance. Motion on the hurricane scale does, however, remain in hydrostatic balance although motion in the individual cumulus towers is not in hydrostatic equilibrium. Contrary to the situation for weak tropical wave disturbances which we have previously discussed, there do exist large-amplitude temperature fluctuations in hurricanes. The magnitude of these temperature fluctuations can be estimated by scaling arguments similar to those used in Section 12.1.

We first require a relationship between the mass field and tangential velocity field for an axially symmetric hurricane vortex. The gradient wind balance (3.11) can be expressed in cylindrical isobaric coordinates as

$$\frac{v^2}{r} + fv = \frac{\partial \Phi}{\partial r} \qquad (12.68)$$

where r is the radial distance from the axis of the storm (positive outward) and v is the tangential velocity (positive for cyclonic flow). We can eliminate Φ in (12.68) with the aid of the hydrostatic equation in the log-pressure coordinate system,

$$\frac{\partial \Phi}{\partial z^*} = \frac{RT}{H}$$

to obtain a relationship between the radial temperature gradient and the vertical shear of the tangential wind:

$$\frac{\partial v}{\partial z^*} \left(f + \frac{2v}{r} \right) = \frac{R}{H} \frac{\partial T}{\partial r} \qquad (12.69)$$

Now the cyclonic flow in a hurricane is observed to decrease quite rapidly with height from its maximum values in the lower troposphere. Thus, above the boundary layer $\partial v / \partial z^* < 0$ which by (12.69) implies that $\partial T / \partial r < 0$ and the temperature maximum must occur at the center of the storm. This is consistent with the observation that hurricanes are *warm core* systems, and of course is necessary if the cumulus heating is to generate kinetic energy.

If we let the vertical scale of the system be equal to the scale height H the tangential velocity scale be $U \sim 50$ m sec^{-1}, the horizontal scale be $L \sim 100$ km, and assume that $f \sim 5 \times 10^{-5}$, we find from (12.69) that the radial temperature fluctuation must have a magnitude

$$\Delta T \sim \frac{LH}{R} \left(\frac{U}{H} \right) \left(f + \frac{2U}{L} \right) \sim 10°C$$

Therefore, the radial advection of temperature cannot be neglected in the thermodynamic energy equation. The approximate expression of the thermodynamic energy equation (12.36) which was valid for weak tropical disturbances cannot be used in modeling the dynamics of a fully developed hurricane.

The origin of tropical cyclones is still a matter of controversy. It is not clear under what conditions a weak tropical disturbance can be transformed into a hurricane. Although there are many tropical disturbances each year, only rarely does one develop into a hurricane. Thus, the development of a hurricane must require rather special conditions. Theoretical investigations of this problem to date have nearly always assumed the existence of a small cylindrically symmetric disturbance, and examined the conditions under which unstable amplification of the disturbance can occur. The results of those investigations show that under favorable conditions of moisture supply, the CISK mechanism can lead to rapid development of a hurricane scale disturbance. In this amplification process friction must be thought of as playing an energy *producing* role. The kinetic energy loss due to frictional convergence in the boundary layer is more than balanced by the energy generated from the latent heating due to the moisture which the frictional convergence supplies to the cumulus. The first study of this type was due to Charney and Eliassen (1964) who used a model with two levels in the vertical. In their model the heating field is related to the large-scale variables by assuming that all moisture which converges into a vertical column is condensed out as rain and that the latent heat released is distributed in the vertical in proportion to the heat release for moist adiabatic ascent. The boundary layer is assumed to be an Ekman layer so that the boundary layer convergence is proportional to the vorticity at the top of the layer. Even such a simplified model results in a rather complicated mathematical analysis. However, Charney and Eliassen did obtain solutions in the form of self-amplifying perturbations. The computed growth rates of the unstable perturbations in their model are shown as a function of the radial scale of the convective region in Fig. 12.12. For an atmosphere with mean relative humidity of 80 percent, which corresponds to observed conditions in regions of hurricane genesis, a doubling time of about one day results for a disturbance with a radial scale of about 100 km. The growth rate curve remains constant for smaller disturbances. Inclusion of viscosity in the model would no doubt damp the growth rate for the smaller scales. Therefore, the conclusion from Charney and Eliassens study is that the cooperative interaction of cumulus convection and the large-scale fields can produce an amplifying disturbance on the hurricane scale provided that the mean atmosphere is unsaturated but conditionally unstable.

Several numerical models to simulate the life cycle of hurricanes have also been developed. All successful models of this type have used some form of the CISK mechanism as the energy source for the hurricane. One of the most interesting conclusions to come from this work is the confirmation of the observational evidence that the hurricane can maintain itself only in the presence of very warm ocean surface temperatures. For surface temperatures less than 26°C, the converging flow in the boundary layer apparently cannot

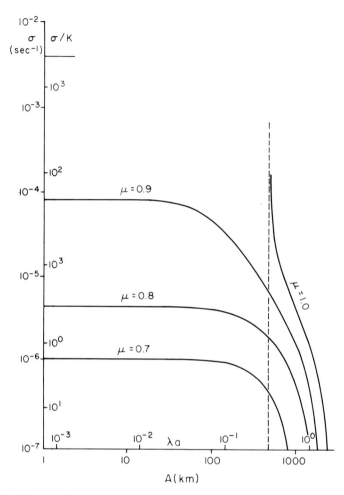

Fig. 12.12   The growth rate σ as a function of the radical scale A of the convective region for various values of the relative humidity μ in the CISK model of Charney and Eliassen (1964.)

acquire a high enough equivalent potential temperature to sustain the intense rate of convective heating needed to maintain the hurricane circulation.

Finally, it should be said that most theoretical studies of the hurricane to date have been limited to axially symmetric models. However, observed precipitation patterns in the hurricane are hardly symmetric. Rather the convection tends to concentrate along *spiral bands* radiating from the center of the storm. It is likely that this asymmetry plays an important role in the evolution of tropical storms and must be included in future models.

Problems

1. An air parcel at 920 mb with temperature 20°C is saturated (mixing ratio 16 gm kg^{-1}). Compute θ_e for the parcel.

2. Suppose an entraining cumulus updraft has a vertical mass flux M which increases exponentially with height according to $M = e^{z/H}$ if the updraft speed is 3 m sec^{-1} at 2-km height, what is its value at a height of 8 km assuming that the updraft has zero net buoyancy?

3. With the aid of (12.50) determine the relationship of the Φ' and $w^{*\prime}$ fields to u' for the Kelvin wave.

4. Verify by direct substitution that the mixed Rossby–gravity mode solution (12.53) indeed satisfies the wave equations (12.41)–(12.44).

5. Consider the following profiles of the mean zonal current:

$$\text{(a)} \qquad \bar{u}(y) = +u_0 \tanh[l(y - y_0)]$$
$$\text{(b)} \qquad \bar{u}(y) = -u_0 \sin^2[l(y - y_0)]$$

where u_0, y_0, and l are constants and y is the distance from the equator. Determine the necessary conditions for each profile to be barotropically unstable.

Suggested References

Riehl, *Tropical Meteorology*, although somewhat out of date remains the basic reference on the descriptive aspects of the subject.

Palmén and Newton, *Atmospheric Circulation Systems*, contains a brief but more up-to-date review of observational aspects of tropical circulations.

Wallace (1971) is a detailed review of the structure of synoptic scale tropospheric wave disturbances in the equatorial Pacific.

Lindzen (1967) discusses the basic theory of large scale waves on the equatorial β plane.

Ogura (1963) briefly reviews theoretical work on the dynamics of cumulus convection. This paper also has a helpful bibliography of original papers on the subject of convection.

Rosenthal (1970) describes one of the several numerical models designed to simulate hurricane development.

Appendix

A | Useful Constants

Gravitational constant	$G = 6.668 \times 10^{-8}$ dyne cm^2 gm^{-2}
Effective gravity (at sea level and 45° latitude)	$g = 981$ cm sec^{-2}
Mean radius of the earth	$a = 6.37 \times 10^8$ cm
Angular speed of rotation of the earth	$\Omega = 7.292 \times 10^{-5}$ sec^{-1}
Gas constant for dry air	$R = 2.87 \times 10^6$ cm^2 sec^{-2} deg^{-1}
Specific heat of air at constant pressure	$c_p = 9.96 \times 10^6$ cm^2 sec^{-2} deg^{-1}
Specific heat of air at constant volume	$c_v = 7.09 \times 10^6$ cm^2 sec^{-2} deg^{-1}

$$\gamma \equiv c_p/c_v = 1.4$$

$$\kappa \equiv R/c_p = 0.288$$

Universal gas constant	$R^* = 8.317 \times 10^7$ erg mol^{-1} deg^{-1}
Molecular weight of water	$m_v = 18.016$ gm mol^{-1}
Latent heat of condensation (at 0°C)	$L_c = 2.5 \times 10^{10}$ ergs gm^{-2}

Appendix

B | List of Symbols

Only the principal symbols are listed. Symbols formed by adding primes or subscripted indices are not listed separately. Boldface type indicates vector quantities. Where symbols have more than one meaning, the sections where the second meaning applies are indicated in the list.

a	(1) Radius of the earth; (2) inner radius of a laboratory annulus (Section 11.15)
b	Outer radius of a laboratory annulus
c	Phase speed of a wave
c_p	Specific heat of dry air at constant pressure
c_{pv}	Specific heat of water vapor at constant pressure
c_v	Specific heat of dry air at constant volume
c_w	Specific heat of liquid water
d	Grid distance
e	Vapor pressure
e_s	Saturation vapor pressure
f	Coriolis parameter, $2\Omega \sin \phi$
g	Magnitude of effective gravity
$\mathbf{g}$	Effective gravity
$\mathbf{g}^*$	Gravitational acceleration
h	Depth of fluid layer
i	Square root of minus one
$\mathbf{i}$	Unit vector along the x axis
$\mathbf{j}$	Unit vector along the y axis
$\mathbf{k}$	Unit vector along the z axis

302

k	Zonal wave number
l	(1) Eddy mixing length: (2) meridional wave number
m	A mass element
m_v	Molecular weight of water
n	Distance in direction normal to a parcel trajectory
$\mathbf{n}$	Unit vector normal to a parcel trajectory
p	Pressure
p_s	(1) Standard constant pressure; (2) surface pressure in σ-coordinate system (Section 8.7)
q	Water vapor mixing ratio
q_s	Saturation mixing ratio
r	Radial distance in spherical coordinates
$\mathbf{r}$	A position vector
s	Distance along a parcel trajectory
t	Time
$\mathbf{t}$	Unit vector parallel to a parcel trajectory
u	x component of velocity (eastward)
v	y component of velocity (northward)
w	z component of velocity (upward)
w^*	Vertical motion in log-pressure system
x, y, z	Eastward, northward, and upward distances, respectively, in spherical coordinates or on a β plane
z^*	$-H \ln(p/p_s)$; vertical coordinate in log-pressure system
$\mathbf{A}$	An arbitrary vector
A	(1) Area; (2) weighting function in the equivalent barotropic model (Section 8.4)
A_M, A_T	Wave amplitudes
A_Z	Eddy mixing coefficient
De	Depth of Ekman layer
E_I	Internal energy
$\mathbf{F}$	A force
$\mathbf{Fr}$	Frictional force
G	Universal gravitational constant
H	Scale height
$\dot{H}$	Diabatic heating rate per unit mass
K	(1) Eddy viscosity coefficient; (2) kinetic energy
L	A length scale
M	Mass
N	Brunt–Vaisalla frequency
P	Available potential energy
Q	Entropy
R	(1) Gas constant for dry air; (2) distance from the axis of rotation of the earth to a point on the surface of the earth
$\mathbf{R}$	Vector in the equatorial plane directed from the axis of rotation to a point on the surface
R^*	Universal gas constant
S	$R\Gamma/H$
T	Temperature
T_0	Standard temperature, constant or dependent only on height
U	Horizontal velocity scale

V Velocity (in Chapters 1–6 V is the total velocity, in Chapters 7–12 V designates the *horizontal* velocity)

V_h Horizontal velocity (Chapter 5, only)

W Vertical motion scale

X Meridional streamfunction

α Specific volume

β (1) df/dy, variation of the Coriolis parameter with latitude; (2) the angular direction of the wind (Section 3.3)

γ c_p/c_v the ratio of the specific heats

ε (1) Rate of frictional energy dissipation; (2) thermal expansion coefficient of water

ζ Vertical component of relative vorticity

θ Potential temperature

θ_e Equivalent potential temperature

κ R/c_p ratio of gas constant to specific heat at constant pressure

λ Longitude, positive eastward

μ (1) Dynamic viscosity coefficient; (2) angular momentum per unit mass

ν (1) Angular frequency of a wave; (2) kinematic viscosity (Section 1.2)

ρ Density

σ (1) $-\alpha\,\partial\theta/\partial p$, static stability parameter in isobaric coordinates; (2) p/p_s, vertical coordinate in σ system (Section 8.7)

τ_z Horizontal frictional stress due to vertical shear

ϕ Latitude

χ Geopotential tendency

ψ Streamfunction

ω Vertical wind component (dp/dt) in isobaric coordinates

Γ $T\,\partial \ln \theta/\partial z^*$, a measure of the lapse rate of potential temperature in the log-pressure system

Φ Geopotential

Ω (1) Angular speed of rotation of the earth; (2) angular speed of rotation of laboratory annulus (Section 8.5)

$\boldsymbol{\Omega}$ Angular velocity of the earth

C Decomposition of a Vector into the Sum of Irrotational and Nondivergent Parts

We wish to show that any vector $\mathbf{V}$ may be written in the form

$$\mathbf{V} = \mathbf{V}_r + \mathbf{V}_e \tag{C.1}$$

where $\mathbf{V}_r$ is a nondivergent vector and $\mathbf{V}_e$ an irrotational vector, that is,

$$\mathbf{\nabla} \times \mathbf{V}_e = 0 \quad \text{and} \quad \mathbf{\nabla} \cdot \mathbf{V}_r = 0$$

To prove (C.1) we define a vector $\mathbf{W}$ such that

$$\nabla^2 \mathbf{W} = -\mathbf{V} \tag{C.2}$$

Then using a vector identity we may write

$$-\nabla^2 \mathbf{W} = -\mathbf{\nabla}(\mathbf{\nabla} \cdot \mathbf{W}) + \mathbf{\nabla} \times \mathbf{\nabla} \times \mathbf{W} \tag{C.3}$$

We next define a *scalar potential* χ and a *vector potential* $\mathbf{A}$ as follows

$$\chi \equiv \mathbf{\nabla} \cdot \mathbf{W}, \quad \mathbf{A} \equiv \mathbf{\nabla} \times \mathbf{W} \tag{C.4}$$

Thus, from (C.2)–(C.4) we find that

$$\mathbf{V} = \mathbf{\nabla}\chi + \mathbf{\nabla} \times \mathbf{A} \tag{C.5}$$

But it can be shown by direct expansion into components that

$$\mathbf{\nabla} \times \mathbf{\nabla}\chi = 0 \quad \text{and} \quad \mathbf{\nabla} \cdot (\mathbf{\nabla} \times \mathbf{A}) = 0$$

thus, $\mathbf{\nabla}\chi = \mathbf{V}_e$ and $\mathbf{\nabla} \times \mathbf{A} = \mathbf{V}_r$ which was to be proved.

If $\mathbf{V}$ is a two-dimensional vector, $\mathbf{V} = \mathbf{i}u + \mathbf{j}v$, then $\mathbf{V} \times \mathbf{A}$ must have its $\mathbf{k}$ component zero. In that case $\mathbf{A} = \mathbf{k}A_z \equiv \mathbf{k}\psi$ and we may write

$$\mathbf{V}_r = \mathbf{V} \times (\mathbf{k}\psi) = \mathbf{k} \times \nabla\psi \qquad (C.6)$$

Thus in the two-dimensional case $\mathbf{V}_r$ is uniquely determined by the *stream-function* ψ. The complete two-dimensional velocity field can thus be written in cartesian components as

$$u = -\frac{\partial\psi}{\partial y} + \frac{\partial\chi}{\partial x}, \qquad v = +\frac{\partial\psi}{\partial x} + \frac{\partial\chi}{\partial y} \qquad (C.7)$$

From which it follows that

$$\zeta = \frac{\partial v}{\partial x} - \frac{\partial u}{\partial y} = \left(\frac{\partial^2}{\partial x^2} + \frac{\partial^2}{\partial y^2}\right)\psi$$

$$\frac{\partial u}{\partial x} + \frac{\partial v}{\partial y} = \left(\frac{\partial^2}{\partial x^2} + \frac{\partial^2}{\partial y^2}\right)\chi$$

Appendix

D | The Equivalent Potential Temperature

A mathematical expression for θ_e can be derived by applying the first law of thermodynamics to a mixture of 1 gm dry air plus q gm of water vapor (q, called the *mixing ratio*, is usually expressed as grams of vapor per kilogram of dry air). If the parcel is not saturated, the dry air satisfies the energy equation

$$c_p \, dT - \frac{d(p - e)}{p - e} RT = 0 \tag{D.1}$$

and the water vapor satisfies

$$c_{pv} \, dT - \frac{de}{e} \frac{R^*}{m_v} T = 0 \tag{D.2}$$

where the motion is assumed to be adiabatic. Here, e is the partial pressure of the water vapor, c_{pv} the specific heat at constant pressure for the vapor, R^* the universal gas constant, and m_v the molecular weight of water. If the parcel is saturated, then condensation of $-dq_s$ gm vapor per gram dry air will heat the mixture of air and vapor by an amount $-L_c \, dq_s$, where L_c is the latent heat of condensation. Thus, neglecting the small amount of heat which goes into the liquid water, the saturated parcel must satisfy the energy equation

$$c_p \, dT + q_s c_{pv} \, dT - \frac{d(p - e_s)}{(p - e_s)} RT - q_s \frac{de_s}{e_s} \frac{R^*}{m_v} T = -L_c \, dq_s \tag{D.3}$$

where q_s and e_s are the saturation mixing ratio and vapor pressure, respectively. The quantity de_s/e_s may be expressed in terms of temperature using the Clausius-Clapeyron equation[1]

$$\frac{de_s}{dT} = \frac{m_v L_c e_s}{R^* T^2} \tag{D.4}$$

Substituting from (D.4) into (D.3) and rearranging terms we obtain

$$L_c d\left(\frac{q_s}{T}\right) = c_p \frac{dT}{T} - \frac{Rd(p - e_s)}{p - e_s} + q_s c_{pv} \frac{dT}{T} \tag{D.5}$$

If we now define the potential temperature of the dry air θ_d, according to

$$c_p \, d \ln \theta_d = c_p \, d \ln T - R \, d \ln(p - e_s)$$

we can rewrite (D.5) as

$$-L_c \, d\left(\frac{q_s}{T}\right) = c_p \, d \ln \theta_d + q_s c_{pv} \, d \ln T \tag{D.6}$$

However, it may be shown that

$$\frac{dL_c}{dT} = c_{pv} - c_w \tag{D.7}$$

where c_w is the specific of liquid water. Using (D.7) to eliminate c_{pv} in (D.6) we obtain

$$-d\left(\frac{L_c q_s}{T}\right) = c_p \, d \ln \theta_d + q_s c_w \, d \ln T \tag{D.8}$$

Neglecting the last term in (D.8) we may integrate from the original state $(p, T, q_s, e_s, \theta_d)$ to a state where $q_s \simeq 0$. The result is

$$c_p \ln\left(\frac{\theta_d}{\theta_e}\right) = -\frac{L_c q_s}{T}$$

where θ_e is the potential temperature in the final state with $q_s \simeq 0$. Therefore, the equivalent potential temperature of a saturated parcel is given by

$$\theta_e = \theta_d \exp(L_c q_s/c_p T) \simeq \theta \exp(L_c q_s/c_p T) \tag{D.9}$$

Equation (D.9) may also be applied to an unsaturated parcel provided that the temperature used is the temperature that the parcel would have if brought to saturation by an adiabatic expansion.

---

[1] For a derivation, see, for example, Hess, *Introduction to Theoretical Meteorology*, p. 46.

Bibliography

Batchelor, G. K. (1967). "An Introduction to Fluid Dynamics." Cambridge Univ. Press, London.

Brown, R. A. (1970). A secondary flow model for the planetary boundary layer. *J. Atmos. Sci.* **27**, 742–757.

Chang, C. P. (1970). Westward propagating cloud patterns in the Tropical Pacific as seen from time-composite satellite photographs. *J. Atmos. Sci.* **27**, 133–138.

Chapman, S., and Lindzen, R. S. (1970), "Atmospheric Tides." Reidel, Dordrecht, Holland.

Charney, J. G. (1947). The dynamics of long waves in a baroclinic westerly current. *J. Meteorol.* **4**, 135–163.

Charney, J. G., and Eliassen, A. (1964). On the growth of the hurricane depression. *J. Atmos. Sci.* **21**, 68–75.

Cole, J. D. (1968). "Perturbation Methods in Applied Mathematics." Ginn (Blaisdell), Waltham, Massachusetts.

Eckart, C. (1960). "Hydrodynamics of Oceans and Atmospheres." Pergamon, New York.

Feynman, R. P., Leighton, R. B., and Sands, M. (1963), "The Feynman Lectures on Physics." Addison-Wesley, Reading, Massachusetts.

Fultz, D. (1961). Developments in controlled experiments on larger scale geophysical problems. *Advan. Geophys.* **7**, 1–103.

Gerbier, N., and Gerenger, M. (1961). Experimental studies of lee waves in the French Alps. *Quart. J. Roy. Meteorol. Soc.* **87**, 13–23.

Greenspan, H. P. (1968). "The Theory of Rotating Fluids." Cambridge Univ. Press, London.

Haltiner, G. J. (1971). "Numerical Weather Prediction." Wiley, New York.

Haltiner, G. J., and Martin, F. L. (1957). "Dynamical and Physical Meteorology," McGraw-Hill, New York.

Hess, S. L. (1959). "Introduction to Theoretical Meteorology." Holt, New York.

Hildebrand, F. B. (1962). "Advanced Calculus for Applications." Prentice-Hall, Englewood Cliffs, New Jersey.

Kittel, C., Knight, W. D., and Ruderman, M. A. (1965). "Berkeley Physics Course." Volume I. McGraw-Hill, New York.

Kornfield, J., Hasler, A. F., Hanson, K. J., and Suomi, V. E. (1967). Photographic cloud climatology from ESSA III and V computer-produced mosaics. *Bull. Amer. Meteorol. Soc.* **48**, 878–882.

Lindzen, R. S. (1967). Planetary waves on beta-planes. *Monthly Weather Rev.* **95**, 441–451.

Lindzen, R. S., Batten, E. S., and Kim, J. W. (1968). Oscillations in atmospheres with tops. *Monthly Weather Rev.* **96**, 133–140.

Lindzen, R. S., and Holton, J. R. (1968). A theory of the quasibiennial oscillation. *J. Atmos. Sci.* **25**, 1095–1107.

Lorenz, E. N. (1960). Energy and numerical weather prediction. *Tellus* **12**, 364–373.

Lorenz, E. N. (1967). "The Nature and Theory of the General Circulation of the Atmosphere." World Meteorological Organization, Geneva, Switzerland.

Lumley, J. L., and Panofsky, H. A. (1964). "The Structure of Atmospheric Turbulence." Wiley (Interscience), New York.

Manabe, S., Smagorinsky, J., and Strickler, R. F. (1965). Simulated climatology of a general circulation model with a hydrologic cycle. *Monthly Weather Rev.* **93**, 769–798.

Matsuno, T. (1966). Quasi-geostrophic motions in the equatorial area. *J. Meteorol. Soc. Japan* **44**, 25–42.

Milne-Thompson, L. M. (1960). "Theoretical Hydrodynamics" (4th ed.). Macmillan, New York.

Newell, R. E. (ed.) (1971). "Observational Aspects of the Tropical General Circulation." MIT Press, Cambridge, Massachusetts.

Nitta, T., and Yanai, M. (1969). A note on barotropic instability of the tropical easterly current. *J. Meteorol. Soc. Japan* **47**, 183–197.

Ogura, Y. (1963). A review of numerical modeling research on small scale convection in the atmosphere. *Meteorol. Monographs*, **5** (27), 65–76.

Oort, A. H. (1964). On estimates of the atmospheric energy cycle. *Monthly Weather Rev.* **92**, 483–493.

Ooyama, K. (1969). Numerical simulation of the life cycle of tropical cyclones. *J. Atmos. Sci.* **26**, 3–40.

Palmen, E., and Newton, C. W. (1969). "Atmospheric Circulation Systems." Academic Press, New York.

Phillips, N. A. (1956). The general circulation of the atmosphere: a numerical experiment. *Quart. J. Roy. Meteorol. Soc.* **82**, 123–164.

Phillips, N. A. (1963). Geostrophic motion. *Rev. Geophys.* **1**, 123–176.

Phillips, N. A. (1970). Models for weather prediction. *Ann. Rev. Fluid Mechan.* **2**, 251–292.

Platzman, G. W. (1967). A retrospective view of Richardson's book on weather prediction. *Bull. Amer. Meteorol. Soc.* **48**, 514–551.

Reed, R. J., and Rogers, D. G. (1962). The circulation of the tropical stratosphere in the years 1954–1960. *J. Atmos. Sci.* **19**, 127–135.

Reed, R. J., Wolfe, J. L., and Nishimoto, H. (1963). A spectral analysis of the energetics of the stratospheric sudden warming of early 1957. *J. Atmos. Sci.* **20**, 256–275.

Richardson, L. F. (1922). "Weather Prediction by Numerical Process." Cambridge Univ. Press, London (Reprinted by Dover, New York).

Richtmeyer, R. D., and Morton, K. W. (1967). "Difference Methods for Initial Value Problems" (2nd ed.). Wiley (Interscience), New York.

Riehl, H. (1954). "Tropical Meteorology," McGraw-Hill, New York.

Riehl, H., and Malkus, J. S. (1958). On the heat balance of the equatorial trough zone. *Geophysica* **6**, 503–538.

Rosenthal, S. L. (1968). Linear analysis of a tropical cyclone model with increased vertical resolution, *Monthly Weather Rev.* **96**, 858–866.

Rosenthal, S. L. (1970). A circularly symmetric primitive equation model of tropical cyclone development containing an explicit water vapor cycle. *Monthly Weather Rev.* **98**, 643–663.

Sawyer, J. S. (1956). The vertical circulation at meteorological fronts and its relation to frontogenesis. *Proc. Roy. Soc. A* **234**, 346–362.

Scorer, R. S. (1958). "Natural Aerodynamics." Pergamon, London.

Shuman, F. G., and Hovermale, J. (1968). An operational six-layer primitive equation model. *J. Appl. Meteorol.* **7**, 525–547.

Sinclair, P. C. (1965). On the rotation of dust devils. *Bull. Amer. Meteorol. Soc.* **46**, 388–391.

Smagorinsky, J. (1967). The role of numerical modeling. *Bull. Amer. Meteorol. Soc.* **48**, 89–93.

Smagorinsky, J. (1969). Problems and promises of deterministic extended range forecasting. *Bull. Amer. Meteorol. Soc.* **50**, 286–312.

Thompson, P. D. (1961). "Numerical Weather Analysis and Prediction." Macmillan, New York.

U. S. Navy. "Arctic Forecast Guide." Navy Weather Research Facility NWRF 16-0462-058, April 1962.

Wallace, J. M. (1971). Spectral studies of tropospheric wave disturbances in the tropical Western Pacific. *Rev. Geophys.* **9**, 557–612.

Wallace, J. M., and Kousky, V. E. (1968). Observational evidence of Kelvin waves in the tropical stratosphere. *J. Atmos. Sci.* **25**, 900–907.

Warsh, K. L., Echternacht, K. L., and Garstang, M. (1971). Structure of near-surface currents east of Barbados, *J. Phys. Oceanog.* **1**, 123–129.

Williams, K. T. (1971). A statistical analysis of satellite-observed trade wind cloud clusters in the western North Pacific. *Atmospheric Science Paper No 161*, Department of Atmospheric Science, Colorado State University, Fort Collins, Colorado.

Williams, R. T., and Plotkin, J. (1968). Quasi-geostrophic frontogenesis. *J. Atmos. Sci.* **25**, 201–206.

Index

312

W

Z

International Geophysics Series

Editor

J. VAN MIEGHEM

Royal Belgian Meteorological Institute
Uccle, Belgium